W9-DJF-928

Low Pressure Boilers

Frederick M. Steingress

American Technical Publishers, Inc.
Alsip, Illinois 60658

Copyright © 1970 by
American Technical Publishers, Inc.

All rights reserved

123456789-70-181716

*No portion of this publication may be reproduced
by any process such as photocopying, recording,
storage in a retrieval system or transmitted by
any means without permission of the publisher.*

Library of Congress Catalog Card No.: 75-99223
ISBN: 0-8269-4400-0

Printed in the United States of America

PREFACE

The purpose of this book is to act as an introduction to stationary engineering. It is intended for the operator of boilers. Although it covers Low Pressure Boilers it breaks the ice for anyone who would like to know more. There is no mystery to boiler room operation; it is common sense.

I would like to thank all my friends for their continued encouragement, Mr. Rahy Paul for his patience, the Technical Editors for all their help, and Mr. Harold J. Frost for his painstaking proofreading and suggestions.

The various manufacturers have been given credit throughout for their cooperation.

FRED M. STEINGRESS

CONTENTS

1. In the eighteenth century, an Englishman noticed steam made the lid of a pot jump up and down.

2. He decided that if he put a pipe in the lid, he could lead the steam to do work such as heating a room.

3. Even more steam could be produced by increasing the heating surface with a wide shallow vessel.

4. The boiler was next improved by placing the source of heat on the inside so that an even greater area was heated.

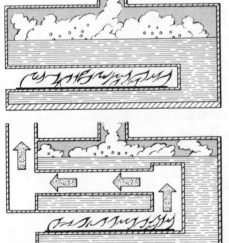

5. To achieve maximum results, coils conducting heat throughout the water produced most steam.

Fig. 1-1. The development of the boiler.

THE BOILER

A boiler is used to produce steam for heat or for power. To generate steam we need three things:

1. Container,
2. Water, and
3. Heat.

The container serves three purposes: it holds the water, it transfers heat to the water to make steam, and it collects the steam that is made. These three functions of the boiler will be discussed in detail as each requires special equipment that the stationary fireman must operate and understand.

Water is used to make steam. We are all familiar with water turning to steam when it is heated. The same thing happens to other liquids although they are used in limited applications. Water is used for steam because it is plentiful and cheap.

Heat is needed to change water to steam. Heat can be supplied in many ways such as heat of the sun, electricity, gas, oil, coal, and wood. These are all used in some cases; however where a large volume of steam is needed as in power and heating uses, a great deal of fuel must be used to provide enough heat. This is done commercially by using fuel oil, coal, or gas.

The stages in the development of a modern boiler are shown in Fig. 1–1. First, we take a container, fill it half full of water, and build a fire under it. When we add a lid to the container we have a simple boiler.

When the fire heats the water to about 212° Fahrenheit the water begins to boil and turn to steam. To use this steam we must collect it and lead it to its working station. We put a pipe on top of the lid of the container so the steam will flow up through the pipe. The pipe leads the steam to where it is needed.

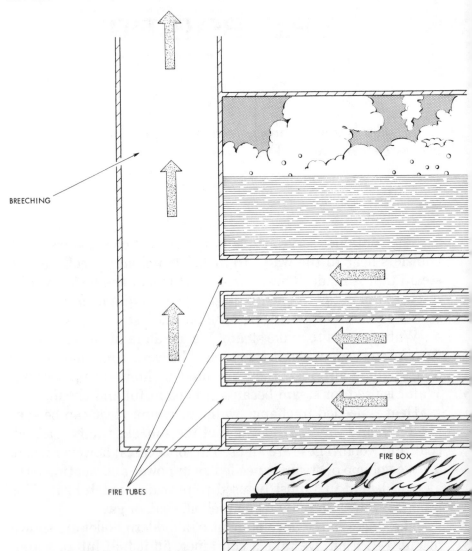

Fig. 1-2. The Scotch marine boiler. Hot gases pass through fire tubes which increase the heating surface where steam can form. The horizontal arrange-

Everything is fine with this hook-up except for a factor that must always be considered—money. We can't waste money making steam. So the boiler must be made more efficient, one that will produce more steam from the same amount of fuel. How can this be done? One way is by giving the heat a larger surface to work on so that more heat will go into the water and produce steam.

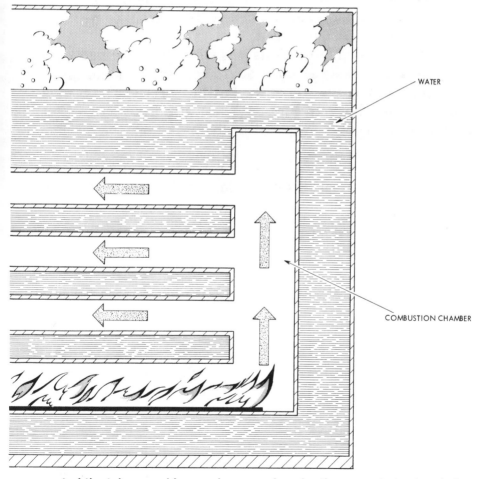

ment of the tubes provides maximum surface for the gases to heat and also slows them so they have more time to heat the water.

We can increase the heating surface (that part of the boiler with water on one side and heat on the other) and put more water closer to the heat by laying the boiler on its side.

This arrangement helps but we can do even better. We can put the fire inside the boiler. If we do this, we get a *combustion chamber* in the boiler. The combustion chamber makes the fuel burn more efficiently because it allows the fuel to mix with a greater quantity of air; the more air that is present, the faster a fuel burns.

This is an improvement, but there is always room for more. The fuel would burn even faster and more completely if it had more air and a larger place to complete its burning. So we make the combustion chamber larger.

Despite these improvements it is still possible to make the boiler work even more economically. The best way to do this is to keep increasing the size of the heating surface. The Scotch Marine Boiler shown in Fig. 1–2, illustrates how the greatest heating surface can be exposed to the heat from the burning fuel.

We still have the same three things we started with—container, water, and heat. The only difference is that we have now put as much water as possible next to the heated metal, resulting in faster and cheaper production of steam.

Although the boiler is a single unit, four separate, independent but interrelated systems are necessary to operate the boiler.

These are basic systems and apply to all boilers regardless of their size or their use.

These systems are:
1. Feed Water System to supply water to the boiler,
2. Fuel System to supply fuel for making heat,
3. Draft System to provide air for combustion,
4. Steam System to collect and control the steam that is made.

Each system is necessary to operate a boiler, yet each is an independent system. Let's look at them one at a time in a general way before taking them up in detail.

FEED WATER SYSTEM

The function of the feed water system is to feed water to the boiler. Fig. 1–3 shows the basic water-to-steam-to-water cycle.

Water in the boiler is heated and turns to steam. The steam leaves the boiler by a pipe called the boiler outlet or main steam line (1) where it enters the *main header* (2). From the main header, *steam mains* (3) carry the steam to *branch lines* and then to the steam *heating equipment* (4). At this point the steam in heating the radiator cools and turns to water called *condensate*. The condensate is separated from the heating equipment by a *trap* (5) which allows condensate but not steam to pass through. The condensate passes along pipes called *condensate return lines* (6) to a *vacuum tank* (7). A *vacuum pump* (8) creates a vacuum that helps to draw the water out of the condensate return lines and into the vacuum tank. The vacuum pump also delivers the water or condensate back to the boiler through pipes called *feed water lines* (9). Once it has returned to the boiler, the water will again be turned into steam and the process will repeat itself.

It is advisable to study this simple water-to-steam-to-water cycle, become familiar with all its parts, and then take a trip to a boiler room and trace out the cycle.

FUEL SYSTEMS

Three fuels are used to heat the water in a boiler. These are *fuel oil, gas,* and *coal.* The fuel system is slightly different for each kind of fuel. In this section we study each system.

Fuel Oil System. You can see in Fig. 1–4 that the starting point of a fuel oil system is the fuel oil tank which is usually buried in the ground. The oil leaves the tank by a pipe called the *suction line*. It then goes to the *fuel oil pump*. The oil leaves the pump under pressure and is forced through more pipes which are called the *fuel oil discharge lines*. From the fuel oil discharge lines some of the oil goes to the *oil burner* where it is

Fig. 1-3. The basic heating system. Water turns to steam, performs its work, turns back into water, and returns to the boiler to repeat the cycle. 1. main steam line; 2. steam header; 3. branch lines; 4. radiator or other heating equipment; 5. steam trap; 6. condensate return lines; 7. vacuum tank; 8. vacuum pump; 9. feed water lines.

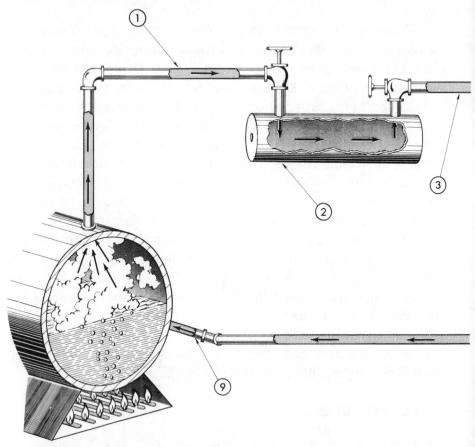

burned. The rest of the oil goes back to the fuel oil tank through the pipes of the *fuel oil return line*.

Gas Fuel System. Boilers use the same gas that you burn at home in cooking and heating. The difference lies in controlling how the gas is burned.

Fig. 1–5 shows a basic gas cycle. Here, city gas lines bring gas to the boiler and release it at city gas pressure. This pressure will vary depending on the city and the particular gas

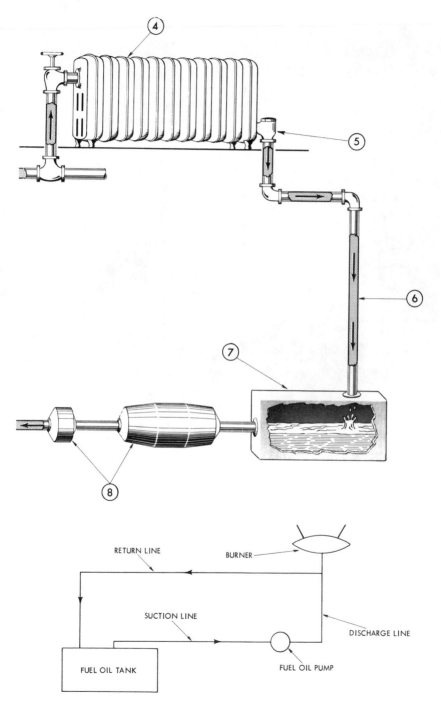

Fig. 1-4. Basic fuel oil system layout. Excess oil returns to tank.

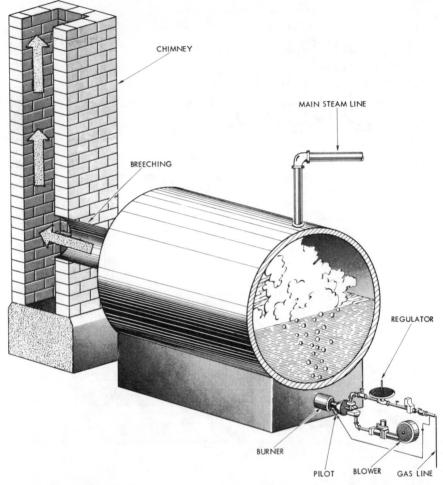

CHIMNEY

MAIN STEAM LINE

BREECHING

REGULATOR

BURNER

PILOT   BLOWER   GAS LINE

Fig. 1-5. Diagram of a gas fuel system. These are increasing in use.

plant. A *regulator* reduces the city gas pressure to zero pounds. Leaving the regulator, the gas mixes with *air* supplied by a *blower*. This mixture then passes into the firebox where it is ignited by a continuously burning pilot flame.

Coal Fuel System. If a boiler uses coal for its fuel, it can be fired either by hand or by stoker. Hand firing is simply shoveling coal into the firebox. This is not a very good method because the fire doors must be left open while the coal is being shoveled. This cools the firebox and brick work some and so reduces efficiency.

Stoker firing, Fig. 1-6, is a mechanical way of putting coal into the firebox. It is more efficient than hand firing.

The fuel you will use will depend on the price and the type of fuel most easily available in your area. At one time, coal was by far the most popular fuel. Now, to a great extent, oil has replaced coal, and gas is becoming popular in many parts of the country.

Fig. 1-6. When coal is used a stoker is more efficient than hand firing. Combustion Engineering Co.

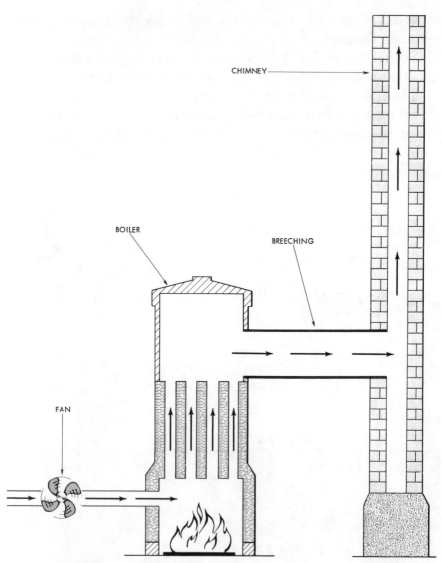

Fig. 1-7. The draft system. The draft supplies air for the fuel to burn.

DRAFT SYSTEM

Before we discuss the draft system, it should be understood that without oxygen nothing can burn. Oxygen comes from the air. So to burn fuel we must allow air to enter the boiler firebox. Not only must we let air into the combustion chamber, we must allow it to leave in the form of gases of combustion after it has mixed with the fuel and burned. Fig. 1–7 shows how the typical draft system works.

A *fan* draws air in and delivers it to the *boiler* under a slight pressure. Here in the boiler firebox it mixes with the fuel and burns. As the air passes through the boiler firebox it enters a pipe called the *breeching*. (This is sometimes called a *smoke pipe* on smaller boilers.) From the breeching the gases of combustion as they are now called, enter a *chimney* or *smoke stack*, and through this they are released into the atmosphere.

STEAM SYSTEM

The only function of a boiler is to produce steam. The steam, once it is produced by the action of heat on water, needs to be controlled and conducted to the places where it is to be used for heat. Because the steam system is so closely involved with the water system, it was described with that system in Fig. 1–3. The main items in the steam system are exclusion of air, control of flow, and maintenance of correct pressure. These are accomplished by means of vents, valves, headers, piping, safety valves, and radiators or heat exchangers.

If all these systems are put together we have a complete boiler system as shown in Fig. 1–3.

Notice that all systems are absolutely necessary to produce steam in the boiler, yet each is separate and independent of the others.

Looking Back

1. Three factors are necessary to produce steam: water, heat, and a suitable container.
2. The container holds water, transfers heat to it, and collects the steam that is produced.
3. Water is supplied to the boiler by a feed water sys-

tem that usually collects condensed steam and returns it.

4. Heat is supplied by burning some fuel in a combustion chamber. Oil, gas, and coal are the fuels used
in low pressure boilers.

5. A draft system is needed to supply the necessary
air to the combustion process.

STEAM BOILER TYPES

So far, we have seen how the boiler produces steam for
heating. While the systems we have described are very general, they apply to all boilers. There are many different types
of boilers and the systems vary somewhat to suit the different
design of each type of boiler. However, the basic principles on
which a boiler operates remains the same.

Like cars, boilers also have different models and styles. Let's
look at some of the different models of boilers.

There are three basic types of steam boilers:

1. The *fire tube* boiler where the heat and gases of combustion pass through tubes that are surrounded by water. Fig.
1–8A,

2. The *water tube* boiler where the water is in the tubes and
the heat and gases of combustion pass among them. Fig. 1–8B,

3. The *cast iron* sectional boiler where the heat and gases
of combustion flow around the sections which contain water.
Fig. 1–8C.

Fire tube boilers consist of several types. One is the *Scotch
marine* boiler which is long, low, and round, Fig. 1–9. The *fire
box* boiler has two different lengths of tubes and much more
brick work than the Scotch marine, Fig. 1–10. The *locomotive*
boiler is another of the fire tube types. It is a stationary model
of the kind used to power steam locomotives, Fig. 1–11. The
vertical fire tube is no longer used in large installations but is
still popular in heating homes, Fig. 1–12.

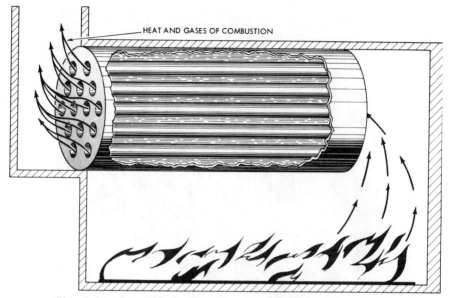

Fig. 1-8A. Fire tube boiler. The heat is inside the tubes.

Fig. 1-8B. Water tube boiler. Water fills the tubes with the fire outside.

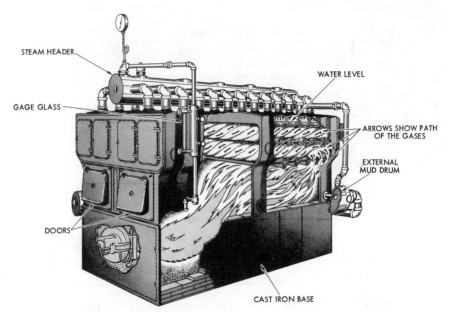

STEAM HEADER

GAGE GLASS

WATER LEVEL

ARROWS SHOW PATH
OF THE GASES

EXTERNAL
MUD DRUM

DOORS

CAST IRON BASE

Fig. 1-8C. Cast iron sectional boiler, water is in large sections.

Fig 1-9. Scotch marine fire tube boiler. Kewanee, American Standard.

Fig. 1-10. A firebox, fire tube boiler.

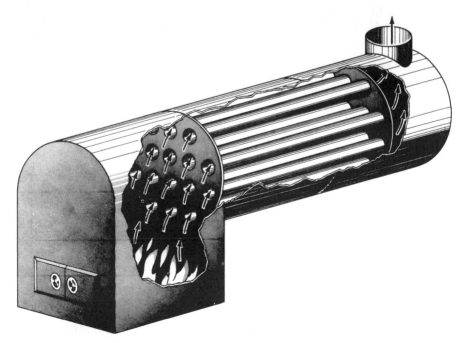

Fig. 1-11. Locomotive type boiler. Derived from the railroad steam engine.

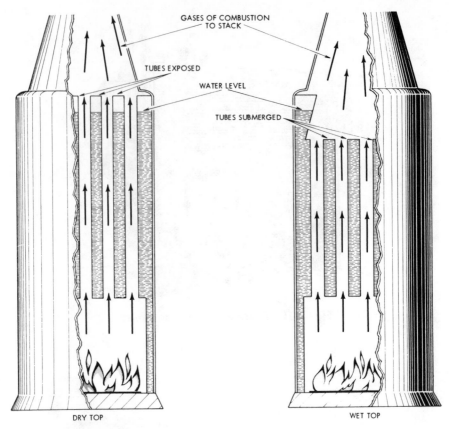

GASES OF COMBUSTION
TO STACK

TUBES EXPOSED

WATER LEVEL

TUBES SUBMERGED

DRY TOP   WET TOP

Fig. 1-12. Two types of vertical fire tube boilers.

In all of these fire tube boilers the gases and heat of combustion pass through tubes that are surrounded by water.

Water tube boilers have many different models such as single drum, multi-drum, straight tube, and bent tube. A straight tube type of water tube boiler is shown in Fig. 1–13. A three-pass water tube boiler is shown in Fig. 1–14. Note that the gases pass over the tubes three times before they leave. This lets the water absorb most heat from the gases of combustion.

In all water tube boilers the name is quite descriptive. The water is inside the tubes while the heat flows around them.

Fig. 1-13. Straight water tube boiler.

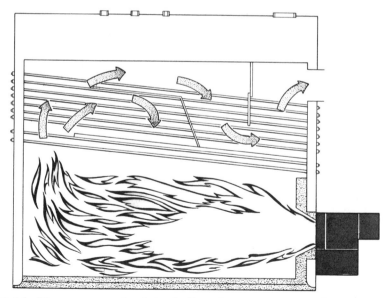

Fig. 1-14. Three pass water tube boiler. Gases pass over the tubes three times. International Boiler Works.

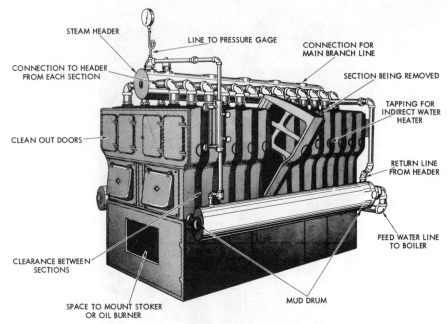

Fig. 1-15. Cast iron sectional boiler. Often called a pork chop boiler. Sections can be added to increase capacity.

Cast iron sectional boilers are the third type. They are sometimes referred to as water tube cast iron boilers, but they really do not have tubes. The water is in cast iron sections which are joined together to form a boiler, Fig. 1-15.

The cast iron sectional may have five sections for a small building and 12 sections for a large one. The sections are fastened together to provide the size needed. They are sometimes called pork chop boilers as the individual sections resemble a pork chop in shape.

Looking Back

1. There are three basic types of boilers. They are fire tube, water tube, and cast iron sectional.

2. Fire tube boilers have the fire's heat and gases of combustion inside the tubes.

3. Water tube boilers have the water inside the tubes and the heat on the outside.

4. Cast iron boilers have large sections with water in them and heat flowing around them.

2

BOILER FITTINGS

Everything on a boiler is there for a definite reason. It is there for safety or efficiency or a combination of both. Whether it is a fire tube, water tube, or cast iron sectional boiler, everything on it is important and necessary.

SAFETY VALVE

In many states the inspectors feel that the safety valve (Fig. 2-1) is the most important valve on a boiler. Let's see why.

Purpose. The maximum allowable working pressure (MAWP) for low pressure boilers is 15 pounds per square inch (psi). That is, the pressure of the steam in the boiler can not go above 15 pounds. The purpose of the safety valve, then, is to prevent the pressure in the boiler from climbing above its MAWP. If the pressure were to rise above the MAWP, the safety valve would pop open and prevent any build-up of pressure that might lead to a boiler explosion.

Location. The safety valve is located at the highest part of the steam side of the boiler and connected directly to the shell. There are *no* valves between the safety and the boiler.

How it works. Safety valves are designed to pop open on pressure and stay open until there is a definite drop in pressure inside of the boiler (this is known as *blow back* or *blow*

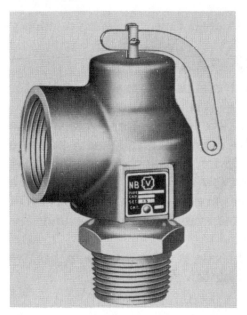

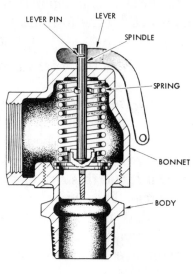

Fig. 2-1. A spring safety valve. The safety valve on a low pressure boiler pops open when the steam pressure inside goes higher than 15 pounds per square inch (psi). Manning, Maxwell and Moore, Inc.

down of safety), and to close without chattering (opening and closing quickly). In most states, the only type of safety valve allowed on steam boilers is the spring loaded type. That is because such a valve is most difficult to tamper with. Fig. 2-1 (right) shows how the spring safety valve is constructed.

We have already seen that the MAWP of a low pressure boiler is 15 pounds per square inch. So, before the safety valve will open, at least 15 pounds of pressure must be exerted against every square inch of the safety valve that is exposed to the steam. In other words, if the area of the safety valve is seven square inches, and if 15 pounds of pressure are pushing against each square inch, then the *total force* of the steam acting against the safety valve will be 105 pounds.

From this we can see that the pressure (in square inches) of steam multiplied by the area of the safety valve will tell us the total force acting against the valve. This can be more simply expressed by the formula:

$$T.F. = P \times A$$

where T.F. is the total force. P stands for pressure, and A for area.

Since the base of a safety valve is in the form of a circle, to find the area of a safety valve exposed to the steam pressure we need only apply the formula for finding the area of a circle. That formula is $D^2 \times 0.7854$. Now 0.7854 is a fixed value or a *constant*, and this remains the same no matter what the size of the circle. The D in the formula stands for diameter and it is squared, or multiplied by itself.

If, then, the diameter of the safety valve is three inches, we can discover its area by applying this formula:

$$\text{Area} = 3 \times 3 \times 0.7854$$

So the total force that 15 pounds of steam pressure would exert on a safety valve whose diameter is three inches would be:

$$
\begin{aligned}
T.F. &= P \times A \\
&= 15 \text{ lbs psi} \times D^2 \times 0.7854 \\
&= 15 \text{ lbs psi} \times (3 \times 3 \times 0.7854) \\
&= 15 \text{ lbs psi} \times 7.0686 \text{ sq in.} \\
&= 106 \text{ lbs.}
\end{aligned}
$$

The spring in the safety valve holds the valve tightly closed against its seat until the steam pressure reaches 15 pounds. Then, as you can see in Fig. 2–2, the total force of the steam slowly lifts the valve off its seat. The steam now enters the *huddling chamber*. Because the huddling chamber is wider, this gives the steam a larger area to push against, so the total force increases. As Fig. 2–2 shows, the total force increases from 106 to 144 pounds and the valve pops open. This relieves the boiler pressure and, by opening quickly, prevents the seat of the safety valve from being damaged by steam.

The valve must stay open until there is a drop in pressure of usually 2 to 4 pounds. Then the safety will close. This pressure drop gives the fireman a chance to get the boiler back under control.

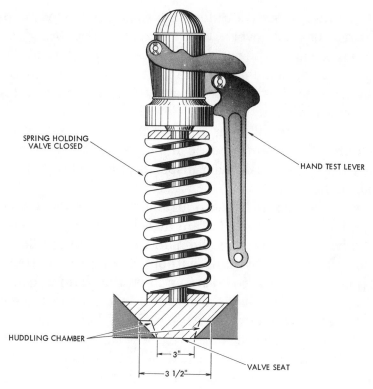

SPRING HOLDING
VALVE CLOSED

HAND TEST LEVER

HUDDLING CHAMBER

←—3"—→

←——3 1/2"——→

VALVE SEAT

Fig. 2-2. Details of a safety valve.

Number of valves. The American Society of Mechanical Engineers (ASME) has set up a code that should be followed if the boiler is to function safely. It specifies the type of material to be used, and the location and number of valves according to the temperature and pressure at which the boiler operates. The ASME Code states that all boilers over 500 sq ft of heating surface (heating surface is any part of a boiler that has water on one side and gases of combustion on the other side) shall have 2 or more safety valves. The pressure on the boiler, then, can not go higher than 6 percent above MAWP with all valves popping.

Testing. Since safety valves are so important, we must be sure they will work when necessary. To be sure of their opera-

tion they must be tested. Safety valves can be tested by hand
(by lifting the test lever), or by pressure (by bringing the
boiler pressure up to the point where the safety will pop). When
testing by hand, make sure there is at least 75 percent of the
popping pressure in the boiler. Safety valves should be tested
at least once a week by hand.

There are many different ideas on how to test safety valves
by pressure. To be sure you do it properly, it would be wise
to check with your local boiler inspector, and follow his recommendations.

Some boiler rooms have an arrangement for testing safety
valves. Chains or wire cables are attached to the handle of a
safety valve. This allows the fireman to stand on the floor and
test the safety valve by hand without having to climb on top
of the boiler. All he has to do is to pull on the chain or cable.

If the plant does have a chain or cable arrangement, it
should be set up to follow the ASME Code. This code has been
adopted in almost all the states. It states that:

"Small chains or wire attached to the levers of pop safety
valves and extended over pulleys to other parts of the boiler
room may be used, but must be arranged so the weight of the
chain or wire will exert no pull on the lever."

When you test a safety valve the worst that can happen is
that it will leak and have to be replaced. But if the valve is not
tested, it may be faulty without your knowing it. Then the
boiler could explode, and that would be much more expensive,
and dangerous, than replacing the safety valve.

Don't take chances. Test your safety valves and test them
often. They are much cheaper to replace than a boiler. Remember, too, that a life can't be replaced at all.

STEAM GAGE

The steam gage is much like the speedometer on your car.
The speedometer shows you how fast you are going and it reads
in miles per hour. The steam gage shows you how much pressure is in your boiler and it reads in pounds per square inch.

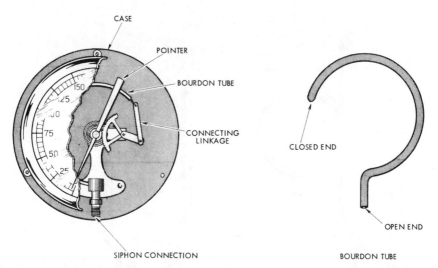

Fig. 2-3. Mechanics of a pressure gage. Pressure causes the Bourdon tube to straighten and moves the pointer on the scale.

The steam gage is also called a *pressure gage.* Fig. 2–3 shows how a pressure gage operates.

Purpose. As we have already seen, low pressure boilers are designed to carry a certain steam pressure (15 psi) which they cannot exceed. The pressure gage allows the fireman to operate his boiler in the safe operating range.

Location. The steam gage must be connected to the highest part of the steam side of your boiler and located so that the fireman can see it.

How it works. Inside the pressure gage there is an oval tube that looks a little like a question mark, Fig. 2–3. This is called a *Bourdon tube* after the French scientist who helped develop the pressure gage.

One end of the Bourdon tube is connected to the steam side of the boiler. The other end is closed and connected to a needle. Fig. 2–3 shows that, as pressure builds up, the Bourdon tube tries to straighten out and moves the needle over a scale to indicate the pressure. But the Bourdon tube is rather delicate

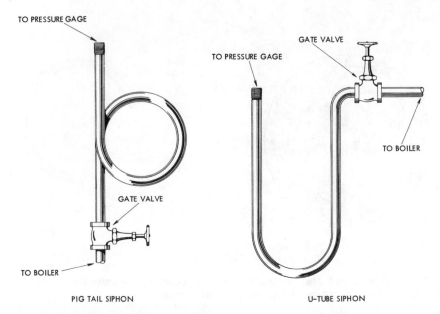

Fig. 2-4. Siphons ensure that only water, not steam, enters the Bourdon tube. This prevents warping of the tube and promotes accuracy.

and the steam entering the tube tends to warp it, and give a false reading. To prevent this, siphons are installed between the boiler and the pressure gage, Fig. 2–4.

The siphon forms a water trap so that water, not steam, enters the Bourdon tube. Never blow water out of the siphon and allow live steam to enter the Bourdon tube or you will damage the pressure gage.

The number on the face of a pressure gage is known as the range of the gage. It should be 1½ to 2 times the MAWP (maximum allowable working pressure) of your boiler, Fig. 2–5 (left).

On low pressure boilers you will usually have a *compound gage*. A compound gage, (Fig. 2–5 right), reads pressure on one side and vacuum on the other.

Vacuum might be defined as an absence of pressure. A vacuum gage is read in inches; for example, a 5 inch vacuum or an

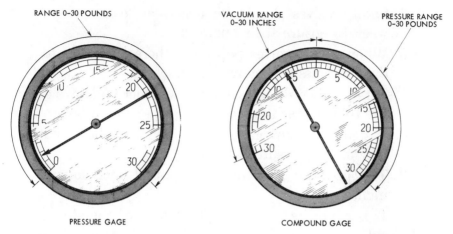

Fig. 2-5. Some common types of pressure gages used on boilers.

8 inch vacuum. A pressure gage is read in pounds per square inch—5 psi or 8 psi.

There are times when a pressure gage through use or misuse gets out of calibration (that is, it no longer reads the proper pressure). This could result in either a fast or a slow gage. A *fast gage* is one that reads more pressure than you actually have in the boiler. A *slow gage* is one that reads less pressure than you have in your boiler. Which do you think would be more dangerous?

Looking Back

1. The safety valve is often considered the most important valve on a boiler.
2. It is connected directly to the shell of the boiler at the highest part of the steam side of the boiler.
3. It prevents the boiler pressure from going above its set pressure.
4. If you have over 500 sq ft of heating surface you must have two or more safety valves.

5. Safety valves should be tested at least once a week to ensure their operation.
6. When safety valves pop open there is a definite drop in pressure before they reseat.
7. Safety valves do the same job whether you have a fire tube, water tube, or cast iron sectional boiler.
8. A steam pressure gage is on a boiler to show you how much pressure is in it.
9. The pressure gage reads in pounds per square inch.
10. It is connected to the highest part of the steam side.
11. A siphon protects the mechanism of the gage from live steam.
12. Compound gages show vacuum on one side and pressure on the other side of the dial.
13. Vacuum is shown in inches.

WATER COLUMN

The ASME Code does not require that all boilers have a water column. Most steam boilers, however, are equipped with one. When the boiler is steaming the water inside is continually in motion and it is difficult to tell just how much is in the boiler.

Purpose. A water column is used to slow down the movement of the water so that you can get an accurate reading of the water level on the gage glass.

At the bottom of the gage glass is a gage glass blowdown valve. This allows us to blow down or clean the gage glass lines of sludge and sediment and to check the water level. All boilers must have two ways of determining the water level. In addition to the gage glass, the try cocks provide the second method of finding the water level in the boiler, Fig. 2–6.

Try cocks can be valves that are opened and closed manually. They can also be weighted valves that are opened by

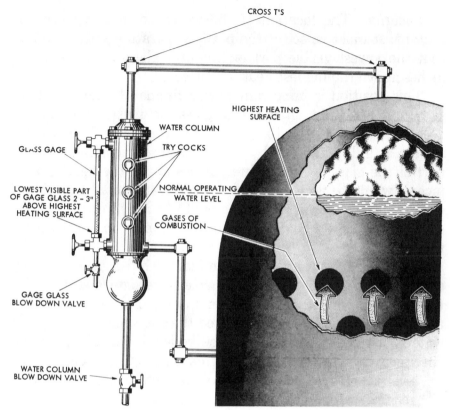

CROSS T'S

HIGHEST HEATING
SURFACE

WATER COLUMN

GLASS GAGE

TRY COCKS

LOWEST VISIBLE PART
OF GAGE GLASS 2 - 3"
ABOVE HIGHEST
HEATING SURFACE

NORMAL OPERATING
WATER LEVEL

GASES OF
COMBUSTION

GAGE GLASS
BLOW DOWN VALVE

WATER COLUMN
BLOW DOWN VALVE

Fig. 2-6. The water column slows down the movement of water in the gage glass and enables the fireman to get an accurate reading of the water level in the boiler.

pulling down on a chain. When released the weight will automatically close the valve.

With a normal operating water level, which is about half a gage glass, water will come out of the bottom try cock when it is opened. The middle try cock will discharge a mixture of steam and water, and the top try cock will yield steam.

The water column blow down valve is used to keep the column and its lines free from sludge and sediment. Both the gage glass blow down valve and the water column blow down valve should be opened every day.

Location. The location of the water column is most important. It must be set at the normal operating water level so that the lowest visible part of the gage glass is two to three inches above the highest heating surface.

This position is important to the fireman because as long as he can see water in the gage glass, no matter how low it goes, he knows it is safe to add water to the boiler. The heating surface is still protected by water because the lowest visible part of the glass is two to three inches above the heating surface. It would be very dangerous to add water to a boiler if you could not see water in the gage glass for it might cause a boiler explosion. This will be discussed in the chapter on boiler operation.

The fireman should be very familiar with the gage glass, try cocks, and water column, and be able to recognize and handle the problems connected with them. For example, if the top line to the gage glass is closed or clogged the gage glass will fill with water. Just the gage glass, not the boiler, is full of water. By checking with the try cocks, the fireman can tell where his water level is. If the bottom line to the gage glass is closed or clogged the water level in the gage glass will remain stationary; it will not move up and down as is normal in a steaming boiler. After a period of time the water in the gage glass will start to get higher due to the steam condensing on top.

BOTTOM BLOW DOWN VALVES

All boilers whether they are fire tube, water tube, or cast iron sectional will have *bottom blow down lines* and on these lines will be *blow down* or *blow off valves*, Fig. 2–7.

Purpose. There are four reasons for using bottom blow down valves. These are:

1. Controlling high water.
2. Removing sludge and sediment.
3. Controlling chemical concentrations in the water.
4. Dumping a boiler for cleaning or inspection.

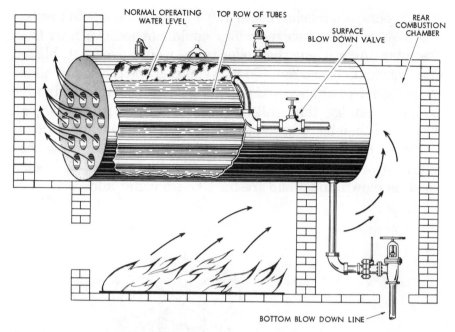

Fig. 2-7. Locations of blow down lines. Blow down lines are at the lowest point of the water side of the boiler.

Dumping. To dump a boiler means to empty it of its contents. It is necessary to do this because boilers should be examined for any defects inside and outside once a year. Before dumping, the boiler has to be taken off the line (shut off from the main header) and cooled. After it has been dumped the boiler should be cleaned thoroughly, both on the fire side and the water side.

High water. There are times when the water level in the boiler gets too high. This could lead to carryover (water being carried over with the steam into the steam lines), which will cause water hammer and possibly a line rupture. To prevent this, when there is high water, the boiler is given a bottom blow until it returns to its normal operating water level.

Control chemicals. All water contains certain minerals or salts. These salts tend to settle out in the form of a scale when the temperature reaches about 150°F. This scale would insulate the heating surface so the water could not remove heat from the tubes. As a result the boiler tubes would overheat, blister, and eventually burn out. To prevent this chemicals are added to the boiler to change scale-forming salts into a non-adhering sludge (a sludge that will stay in suspension in the water), which can then be removed from the boiler by blowing it down.

When these chemicals are added to the boiler, in time they build up or are too concentrated. To dilute them the boiler must be blown down and fresh make-up water added.

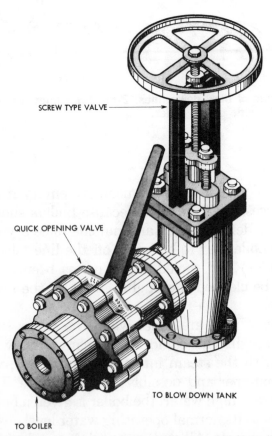

SCREW TYPE VALVE

QUICK OPENING VALVE

TO BLOW DOWN TANK

TO BOILER

Fig. 2-8. Blow down valves. The quick opening valve is for stand-by use. It is closest to the boiler.

Sludge and Sediment. As mentioned above, if chemicals turn the scale forming salts into a sludge that stays in suspension, it must be removed. This is done by giving the boiler a bottom blow. The best time to blow down a boiler to remove sludge and sediment is when the boiler is under a light load.

The water circulation is then slower and the sludge will tend to settle to the bottom of the boiler making it much easier to remove. The discharge from the bottom blow down line *should not* go directly into a city sewer line. It should go first to either a blown down tank or a sump and from there to the sewer. This prevents damage to the sewer.

Location. The bottom blow down line is at the lowest point of the water side of a boiler. On a water tube boiler, there may be one or two blow down valves. When two valves are used, one is usually quick-closing and the other a screw type, Fig. 2–8.

The quick closing valve is a lever operated gate valve that is used as a sealing valve. The screw type valve is used as the blowing valve and it takes all the wear during blow down. It should be constructed so that no pockets of sludge can build up under the seat to prevent it from closing tight. When blowing down a boiler the quick closing valve should *always* be *opened first* and *closed last*.

The quick closing valve is closest to the boiler, and the screw valve is next to it.

SURFACE BLOW OFF LINE

Some boilers have a surface blow off line, Fig. 2–7. This line is located at the normal operating water level and is used to skim off any impurities floating on the surface of the water. This is important because if a scum builds up on the surface, it prevents the steam bubbles from breaking through.

A high surface tension will lead to foaming (that is a rapid fluctuation in water level). One minute the gage glass is full of water, the next minute it is empty. Foaming can in turn cause priming and carryover. By giving the boiler a good surface blow these impurities are removed and foaming reduced.

Looking Back

1. The water column is located at the normal operating water level so that the lowest visible part of the gage glass is two to three inches above the highest heating surface.

2. The top line of the water column connects to the highest part of the steam side of the boiler. The bottom line connects to the water side well below the normal operating water level.

3. The gage glass is located on the water column and provides one way of finding the water level.

4. Two or three try cocks are found on the water column and are another way of finding the water level.

5. If the top line to the gage glass is closed or clogged, the glass will be full of water.

6. If the bottom line to the gage glass is closed or clogged, the water level will remain stationary, slowly filling up.

7. With a normal operating water level the bottom try cock when opened will have water come out, the middle try cock will have a mixture of water and steam, and the top try cock will have only steam.

8. All boilers have blow down lines and blow down valves.

9. The blow down line is always connected to the lowest part of the water side of the boiler.

10. There are four reasons for using the bottom blow down line: to control high water, to remove sludge and sediment, to control chemical concentration of the water, and to dump or drain the boiler.

11. The bottom blow down line should discharge first to a flash tank or open sump and then to the sewer.

12. A surface blow off line is located at the normal operating water level and is used to remove scum and prevent foaming.

FUSIBLE PLUG

Although the ASME Code now only requires them on coal fired boilers, fusible plugs may still be found on gas and oil fired boilers.

Purpose. A fusible plug, Fig. 2-9, is the last warning a fireman has of low water before he starts burning tubes. It is a brass or bronze plug with a tapered, hollow center that is filled with 100 percent pure banka tin which melts at about 450°F.

It can be either a fireside plug, which means it screws from

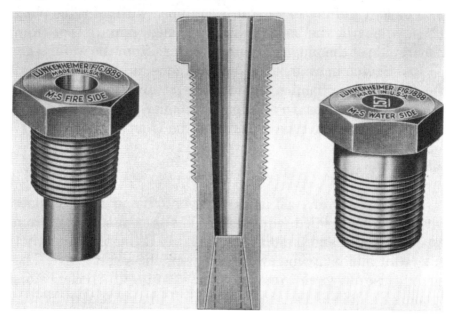

Fig. 2-9. Fire side and water side fusible plugs. When the temperature reaches 450° the tin in the plug melts. The steam rushes through the plug making a whistling sound. Lunkenheimer Co.

the fire side into the water side of the boiler, or it can be a water side plug, which means it screws from the water side into the fire side.

Location. The fusible plug is located one to two inches above the highest heating surface in the direct path of the gases of combustion.

How It Works. As long as water is in contact with the plug it keeps it cool enough to prevent the tin from melting. If, however, the water level gets low, the tin will melt. Steam will then whistle through the plug, warning the fireman.

Fusible plugs must be kept clean both fireside and waterside. If scale were to build up, it could, as explained earlier, prevent the water from removing the heat. In that case, the tin in the plug might melt (this is called *dropping a plug*) even though the water level was normal. Plugs must be replaced annually during inspection. That is because the tin has a tendency to crystallize and this would increase its melting point. It would then be worthless as a warning device. Check your boiler carefully to see if it is equipped with a fusible plug. Then, whenever the boiler is being cleaned, remove soot from the fire side of the plug and scrape all scale from the water side.

In a Scotch marine boiler the plug generally is in the top of the combustion chamber or in the rear tube sheet above the highest row of tubes. In a locomotive boiler, it is the rear tube sheet. In a fire box type boiler it is the front tube sheet.

AIR COCK

Purpose. An air cock is used to vent the air from a boiler when it is being filled with water. This prevents the air from getting trapped and building up pressure. If the pressure is not relieved it builds up and eventually pops the safety valve. The air cock also allows air to escape when warming up a boiler before opening the boiler steam stop valve to the main header. (This is called *cutting a boiler in on the line.*) Finally, it prevents a vacuum from forming in the boiler when taking it out of service (taking it off the line).

The danger of vacuum in a boiler should be clearly understood. Steam is a water vapor in a semi-gaseous condition. It occupies space. When a boiler that has been steaming is being taken off the line, the main boiler steam stop valve is closed. The steam in the boiler cools and condenses thereby creating a vacuum. At sea level the atmospheric pressure is 14.7 psi. If the boiler is under a vacuum it is being subjected to a crushing force of 14.7 pounds on every square inch of surface.

A simple experiment often used in science classes illustrates the force of this pressure. A five gallon can with a small amount of water in it is heated until the water begins to steam. The hole in the can is then sealed and the can is placed under cold water. The steam condenses forming a vacuum and the atmospheric pressure crushes the can.

You can see then that if you allow a vacuum to form on a boiler you are subjecting it to unnecessary strains. If you were to try to remove a hand hole plate (used for cleaning and washing out the boiler) at the bottom of the water side of the boiler with vacuum on the boiler, the plate could be pulled into the water side. The same could happen with the manhole cover located at the top of the steam side of the boiler. Since the manhole cover weighs approximately fifty pounds, it could cause damage to the inside of the boiler. In high pressure plants there have been cases of firemen being pulled into the steam and water drum while trying to remove a manhole cover on a water tube boiler under vacuum.

Never attempt to open the steam and waterside of a boiler until:

1. You are sure there is no steam on the boiler.

2. It is cool enough to dump (empty using the bottom blow down lines).

3. You are sure the air cock is open and there is no vacuum on the boiler.

Location. The air cock is a ½″ or ¾″ line with a valve on it coming off the highest part of the steam side of the boiler. Not all boilers are equipped with an air cock. As a result, the fireman must use the try cocks as vents. Never use a safety valve to vent the boiler. This is often done and can result in damage to the safety valve.

It is very important to vent the air from the boiler when warming it up before cutting it in on the line. If you don't, the air trapped above the water becomes heated, expands, and builds up in pressure. This is air pressure not steam pressure and as a result you will cut the boiler in on the line before it is ready to take its share of the steam load. Then, as soon as you open the boiler's main steam stop valve, the air pressure will drop and the steam from the main header will rush in to take its place. There will be a drop in steam pressure throughout the whole system.

PRESSURETROLS

The pressuretrol, Fig. 2–10, is nothing more than a switch that turns the boiler on and off.

Purpose. The pressuretrol starts and stops the boiler on pressure; it also controls its operating range. For example, an average operating range may be 3 to 6 pounds. When the steam pressure in the boiler drops 3 pounds, the boiler will start up. When the pressure reaches 6 pounds, it will shut off.

Location. The pressuretrol is located at the highest part of the steam side of the boiler. It is important that it be protected with a siphon just as the pressure gage is and that it be mounted straight. Some pressuretrols have a little, arrow-like plumb bob for this purpose. This is especially true of the mercury type of pressuretrol. Unless it is in a vertical position it will not be accurate. Fig. 2–11 shows how the pressuretrol and siphon should be installed.

How it works. The pressuretrol has two adjusting screws and two scales. One scale is the cut-in; the other is the differential. To set a pressuretrol for the operating range you want, you would proceed as follows. The cut-in plus the differential is equal to the cut-out. For example, if you want the boiler to cut in at three pounds and cut out at six pounds, set the cut-in

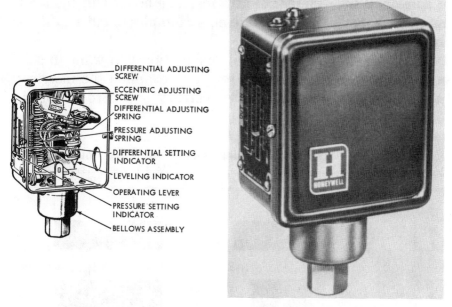

DIFFERENTIAL ADJUSTING SCREW

ECCENTRIC ADJUSTING SCREW

DIFFERENTIAL ADJUSTING SPRING

PRESSURE ADJUSTING SPRING

DIFFERENTIAL SETTING INDICATOR

LEVELING INDICATOR

OPERATING LEVER

PRESSURE SETTING INDICATOR

BELLOWS ASSEMBLY

Fig. 2-10. A pressuretrol turns the boiler on and off.

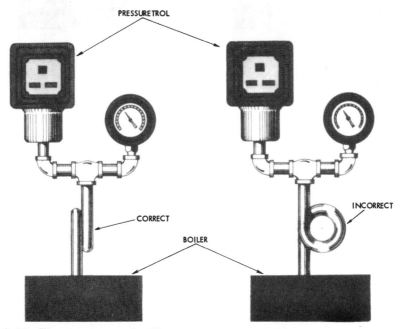

PRESSURETROL

INCORRECT

CORRECT

BOILER

Fig. 2-11. The pressuretrol will not function properly unless a siphon is installed correctly.

at three and the differential at three. The boiler will then cut out at six pounds. Table 2–1 presents examples of cut-in, differential, and cut-out points.

You can get just about any combination you want to suit your needs. It is important to regularly examine the mercury tube in the pressuretrol. If the mercury starts to vaporize,

Table 2-1. The cut-in and differential affect the cut-out.

CUT–IN SETTING +	DIFFERENTIAL SETTING =	CUT–OUT POINT

0   2   4   6   0   2   4   6   0   2   4   6

CUT–IN + DIFFERENTIAL = CUT–OUT

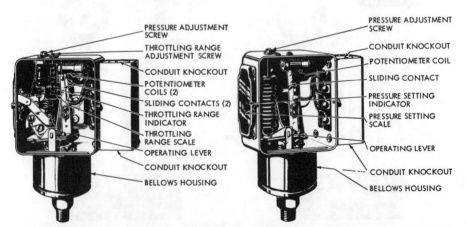

Fig. 2-12. The modulating pressuretrol differs from the ordinary pressure-trol because it controls both high and low fire.

there will not be a good electric contact through the mercury. This will raise the heat. As the heat increases, the mercury will vaporize further, pressure will build up, and the mercury tube will explode. If the mercury tube shows signs of discoloring or the mercury seems to be sticking to the glass in little drops, it would be wise to replace the tube.

A modulating pressuretrol, Fig. 2–12, differs from the pressuretrol just discussed in that it controls high and low fire. High fire is the maximum amount of fuel burned; low fire the minimum.

A boiler should always start off in low fire and shut off in low fire. It should always be firing for longer periods than it is off. This helps to maintain a good furnace temperature, and reduces the cooling effect on brick work, which in turn helps to cut down on maintenance and improves boiler efficiency.

Looking Back

1. The fusible plug is usually found only on boilers burning coal. But it can also be used on boilers burning oil or gas.
2. It is the last warning of low water before a boiler starts burning out tubes.
3. The plug is 1″ to 2″ above the highest heating surface. It melts at about 450° F.
4. It must be kept clean on both fireside and waterside and replaced annually.
5. The air cock is located at the highest part of the steam side of the boiler.
6. The air cock must be kept open when filling a boiler with water to relieve the air pressure.
7. The air cock must be left open to vent the air from the steam side of the boiler when warming it up.
8. The air cock must be open when taking a boiler off the line to prevent a vacuum from forming.

9. Make sure the air cock is open and there is no vacuum in the boiler when you remove a manhole cover or a handhole plate.

10. A pressuretrol is an on and off switch. It starts and stops a boiler on pressure.

11. To set the operating range check your pressuretrol. The cut-in point plus the differential will equal the cut-out.

12. The mercury tube in a pressuretrol should be checked at least twice a year for signs of breakdown of the mercury.

13. The modulating pressuretrol controls high and low fire. High fire is burning maximum fuel; low fire is burning minimum fuel.

14. A boiler should always light off (start up) in low fire and should shut down in low fire.

15. A boiler should always run for longer periods than it is off when burning oil or gas. This helps maintain furnace temperature and reduces cooling of the brick work.

3

FEED WATER ACCESSORIES

Newspapers often carry accounts of violent boiler explosions which kill and maim scores of people. In almost every case, if the boiler had been properly tended, the explosion might have been prevented. As we discuss feed water accessories, then, it is very important to remember one thing: *the fireman is responsible for the safe operation of his boiler.* He should constantly watch the water level in the boiler, and make sure that it does not get too high nor too low. High water can force water into the steam line, which results in water hammer and, possibly, a line rupture. Low water can burn out the boiler tubes and heating surface. More than that, it can cause a boiler explosion. In the interests of safety, the fireman should know every possible way of getting water to his boiler. A thorough knowledge of feed water accessories will enable him to do this.

Before discussing feed water accessories, it would be wise to review the basic water-to-steam-to-water cycle, Fig. 3–1.

The water in the boiler is heated; it turns to steam and leaves the boiler through the main steam line (1) and goes to the main header (2). Here, the mains and branch lines (3) lead from the boiler room and connect with risers which take the steam up and distribute it to the various heating equipment and radiators (4). At this point, the steam, in giving up its heat to the radiator, turns to water. This water is called condensate. The traps (5) after each radiator allow the condensate

Fig. 3-1. The basic boiler system. See Fig. 1-3 if you do not know the parts.

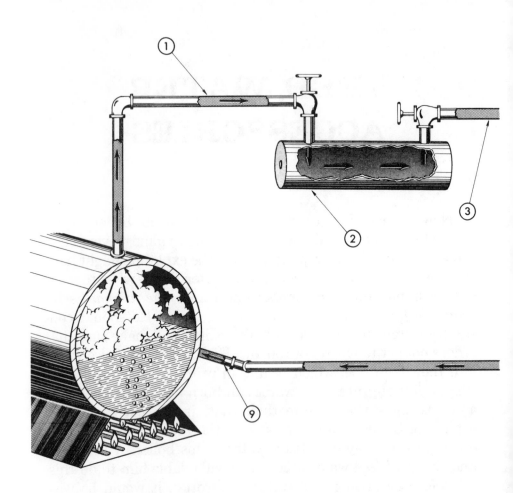

to pass through but not the steam. The condensate is gathered in condensate return lines (6) and is brought back to the vacuum tank (7). The vacuum pump (8) then discharges the water back to the boiler through the feed water lines (9).

We can now examine the various valves and other accessories that help the feed water system function efficiently.

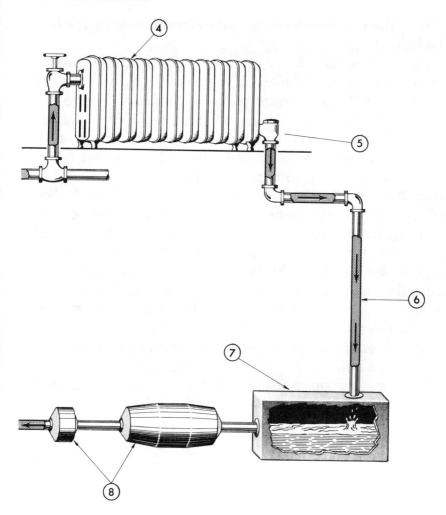

FEED WATER VALVES

Purpose. The feed water line has stop and check valves that work together. The feed water stop valve lets water into the boiler, and the feed water check valve prevents the water from backing out of the boiler into the feed water lines. Figure

3–2 shows these valves and where they are located in relation to the boiler.

Location. The feed water stop valve should be as close to the boiler as practical. The check valve should be between the stop valve and the feed water pump. A glance at Figure 3–2 shows exactly how these valves are arranged. They are arranged so that if the check valve hangs up (fails to open or close), it would be possible to close the stop valve and repair the check valve without dumping the boiler.

How they work. The *feed water stop valve* is controlled manually. The fireman opens and closes it by screwing the stem in or out by hand. Usually, it is a *globe valve.* This has a metal disk in the valve that controls the amount of water flowing through it, see Fig. 3–2 right. When the valve is open, the disk impedes the water passing through the valve, in this way restricting its flow. Globe valves are installed when it is desirable to control the speed at which water flows through a valve.

The *feed water check valve,* on the other hand, is automatic. Usually, it is a swing check valve, as in Fig. 3–2 left. This type of valve opens and closes itself according to the pressure that acts on it. It is designed to let water through only one way.

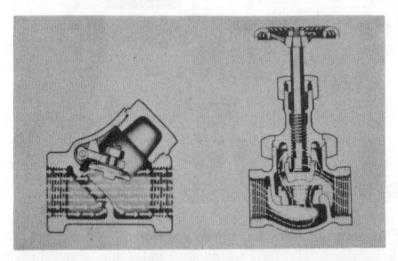

Fig. 3-2. The check valve (left) and the stop valve (right) allow water to feed into the boiler and prevent it from flowing back into the water lines. Kennedy Valve Mfg. Co.

When feed water going to the boiler comes into contact with the valve, it swings open. But if the water tries to flow in the opposite direction (from the boiler into the feed water lines) the valve swings shut.

When the vacuum pump starts to feed water to the boiler, it builds up water pressure in the feed water lines. This pressure overcomes the boiler pressure and the water enters the boiler, swinging open the check valve and passing under the seat of the feed water stop valve. When enough water has been fed into the boiler, the pump stops. Accordingly, the boiler pressure becomes greater than the pressure in the feed water lines. This greater pressure, acting in a reverse direction, closes the check valve, and prevents boiler water from backing up into the feed water lines.

Looking Back

1. To ensure safe operation, the fireman must know every possible way of getting water into the boiler.
2. Water returning to the vacuum pump is called condensate.
3. On the feed water line, the stop valve is closest to the boiler. Next to it is the check valve.
4. The stop valve lets water into the boiler; the check valve prevents water from leaving the boiler.
5. If the check valve sticks open or closed, the fireman can close the stop valve manually. In this way, he can repair the check valve without dumping the boiler.

VACUUM PUMP

Purpose. The vacuum pump serves three purposes:

1. It creates a vacuum on the return lines, drawing back condensate to the vacuum tank;

2. It discharges all the air to the atmosphere;

3. It discharges all the water back to the boiler.

This is the first time we have mentioned air in the return lines. You might wonder where the air comes from if we are supposed to have steam in the lines. Well, water absorbs some air. A certain amount of this is released when the water boils or turns into steam. The released air tends to stay in the lines. But since air inside a boiler can rust, corrode, and pit the boiler metal, it must be removed. So the vacuum pump (Fig. 3–3) discharges it into the atmosphere.

You might like to prove to yourself that warm water contains less air than cold water. Take a kettle of water and boil it. After the water has boiled, pour it into a bottle, and put a cork in the bottle. Wait until the water cools, then drink some. It's flat and tasteless.

Now take the boiled water and pour it back and forth between two glasses. Hold it high when pouring so it will come into contact with as much air as possible. Try drinking it again. It tastes much better now because you have replaced some of the air that was boiled out of it. In the same way, when water in the boiler is heated, it releases its air. As we've already mentioned, we must get rid of this air, and the vacuum pump does that.

How it works. The vacuum pump usually has a switch (Fig. 3–3 (upper inset) that can be set in three positions.

1. *Continuous position:* the pump will run continuously. This position is used for testing the pump.

2. *Float only position:* the pump will start only when the return tank starts to fill up.

3. *Float or vacuum position:* the pump will be started either by a float when the tank becomes full, or on vacuum when the vacuum falls below a predetermined set pressure.

Now we can discuss more fully each of these positions. First, when he wants to test the vacuum pump, the fireman throws the vacuum switch into the continuous position.

Next, the pump is in the float or vacuum position during the heating season. To understand how it works in this position, it

Fig. 3-3. A vacuum pump. Small views show pump construction and the three positions of the selector switch. Nash Engineering Co.

is necessary to know a little about the vacuum control switch. This switch is similar to the pressuretrol, which we have already discussed. The range on the vacuum switch is usually 2″ to 8″. This means that when the vacuum drops to 2″ the pump will start (regardless whether there is water in the vacuum tank; when the vacuum reaches 8″ the pump will shut off. This vacuum helps to pull back the condensate return in the lines.

But, if the vacuum is holding (that is, it is steady within the operating range) and the condensate is coming back, the tank will start to flood. In this case, the float will lift with the rising water level. So the pump will start not on vacuum but on float. This will discharge the accumulated water into the boiler.

In mild weather when the boiler is shut off at night, the switch on the pump is put on *float only*. Here, if some returns are still coming back to the tank, they will be pumped back to the boiler. There is no need for vacuum now.

It is important to remember that the vacuum pump can only return to the boiler the condensate it receives. It takes one pound of water to make one pound of steam. If there is any loss due to leaks the vacuum pump cannot make up this loss. It can only deliver the condensate that comes into it. This is an important point to remember.

CITY WATER MAKE-UP

Purpose. Since the vacuum pump can only put back into the boiler the condensate it receives, there must be some way to add extra water to the boiler to replace the water that has been lost through leaks or by blowing down the boiler. This water is called *make-up water*. It is added through the city water make-up either manually or automatically.

Manual. Some boilers have both a manual and an automatic city water make-up, others have only a manual. If a boiler has both, then, as Fig. 3–4 shows, the manual make-up system simply by-passes the automatic. When he sees that the water level is low, the fireman opens the manual city water make-up valve and city water flows directly into the boiler.

Automatic. The automatic city water make-up valve is a little below the normal operating water level. The top line connects to the top of the steam space; the bottom line to the water side of the boiler well below the normal water level, Fig. 3–4.

The automatic city water make-up is controlled by a float. This float connects to a small needle valve in the city water line. If the water level drops in the boiler, the float drops. This opens the valve in the city water line and feeds water into the boiler. As the water level builds up in the boiler, the float rises and shuts off the automatic city water make-up valve.

Though it lets water into the boiler, the automatic city

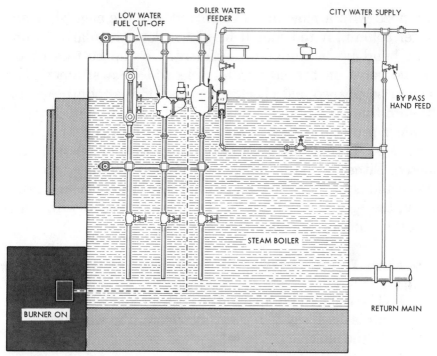

Fig. 3-4. The automatic city water supply feeds the boiler with the water it needs if there is too little condensate being returned.

water make-up is not meant to act as a regulator. It is only meant to make up water that has been lost. If the fireman finds that this valve is feeding water to the boiler at regular intervals, he should check to find out why. It means that he is not getting back the condensate returns. As a result, the make-up valve is adding water to the boiler. The city water make-up contains scale-forming salts which will affect the boiler heating surface unless treated chemically. This is expensive. There is another reason for using as little city water as possible. The returns coming back are warm and relatively free of air. The city water, on the other hand, is cold and contains more air. Excessive use of city water which contains plenty of air would reduce overall efficiency.

Maintenance. The automatic city water make-up is

equipped with a blow down valve and, like the gage glass and water column, it too should be blown down regularly to prevent build up of sludge or sediment. On the bottom of the needle valve on the city water line will be a strainer. This should be cleaned and checked at least once a month or more often if there is any indication of a rapid build-up of dirt. The strainer protects the valve and seat from particles of scale or lime deposits that might cause the valve to stick open or stick closed. Either would be dangerous. If the valve failed to open, you could not get any make-up water into the boiler; if it stuck open, the boiler and all the lines would be flooded with water and the whole system would become waterlogged.

Looking Back

1. The vacuum pump handles both air and water. The air is discharged to the atmosphere; the water goes to the boiler.
2. During the heating season the vacuum pump runs on the float or vacuum position.
3. The vacuum pump can only put back into the boiler the condensates it receives.
4. City water make-up is used to make up water in the system that has been lost due to leaks, blowing down the boiler, or failure to pull back enough condensate returns into the boiler.
5. Make-up water can be added by hand or by an automatic city water make-up valve.
6. If the fireman finds that the city water make-up is being used too often, he must check for leaks in the steam and return lines.
7. The automatic city water make-up valve must be blown down at least once a week to prevent build-up of sludge or sediment that might clog it.
8. The strainer on the city water line before the make-up valve should be cleaned at least once a month.

LOW WATER CUT-OFFS

Purpose. Even though a boiler may be equipped with an automatic city water make-up (and not all boilers are so equipped), it is possible that the automatic city water make-up might *hang up* (fail to work when needed), or there could be a failure of city water. If this were to happen, it could lead to a burned out boiler or a boiler explosion. To protect it, the boiler is equipped with a low water cut-off. The low water cut-off will shut off the burner in the event of low water.

Location. The low water cut-off is a little below the normal operating water level. The top line connects to the highest part of the steam side of the boiler; the bottom line connects to the waterside, well below the normal operating water level. It is equipped with a blow down line to keep the float chamber free of sludge and sediment. There are many models of low water cut-offs. As a rule, smaller boilers use small controls, larger boilers bigger and more intricate controls. The boiler manufacturer usually provides the proper control. Some typical low water cut-offs are shown in Fig. 3–5.

How it works. Despite the many different kinds, the operation of all low water cut-offs is similar. As the water level drops to what would be considered an unsafe level, the float in the low water cut-off drops, and breaks or opens the electric circuit, shutting off the burner.

Fig. 3–6 illustrates a boiler with a normal operating water level. The burner is on, or firing. In Fig. 3–7, the boiler water level has dropped to an unsafe level, and the burner has shut off, thus preventing damage to the boiler. Note that water is still visible in the gage glass when the burner shuts off.

Maintenance. The low water cut-off should be tested daily. The first thing you as a fireman should do upon entering a boiler room is to check the water level of all boilers on the line. This is done by blowing down the gage glass, then the water column, and then the low water cut-off. The boiler should be firing when you blow down the low water-cut-off, and it should shut off as you blow it down.

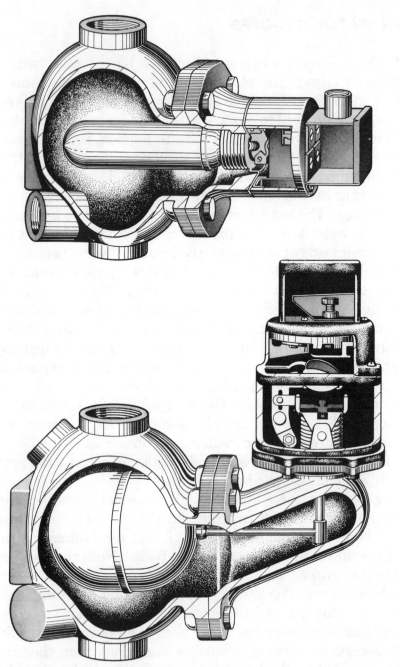

Fig. 3-5. Two types of low water cut-off controls which stop the burner when the water level falls too low. This prevents damage to the heating surfaces and eliminates possibility of an explosion.

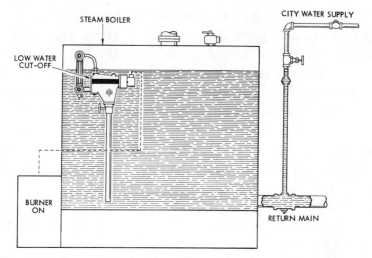

Fig. 3-6. When water level is normal, the burner is on or firing.

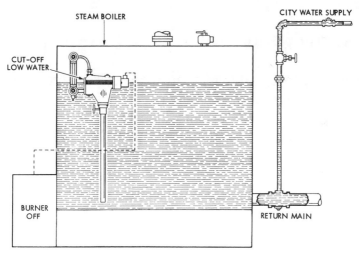

Fig. 3-7. When water drops to an unsafe level, the burner shuts off.

At least once a month, you should test the low water cut-off by allowing the water level to drop in the boiler. This can be done by first shutting off the automatic city water make-up (if the boiler has one) and shutting off the vacuum pump.

Then stand by and make sure the burner shuts off at the proper level in the gage glass. Usually, there are about two inches of water left in the gage glass when the burner is shut off by the low water cut-off. You can then start up the vacuum pump and when the tank is empty, open your automatic city water make-up. This will be discussed again in a later chapter.

Needless to say, the fireman should never walk away from the boiler during this test. He must stand by and watch the water level because if the low water cut-off fails to shut off the boiler . . . well, there just won't be a boiler left.

Some states have rather strongly recommended that boilers be equipped with two low water cut-offs. The second should be installed with separate piping and set at a level just a little below the first. This gives a little added protection. If a boiler is equipped with two low water cut-offs, they should both be tested.

FEED WATER REGULATORS

Purpose. A feed water regulator is a device that maintains a constant water level in the boiler. But before we can discuss this regulator we must stop and look at another type of feed water system. Up to this point, we said that the condensate returned to a vacuum tank, and from there to the boiler, Fig. 3–8. That is the very basic system. In a slightly more complex system the condensate, after leaving the vacuum tank, flows into a *condensate return tank*, and from there into the boiler.

Fig. 3–9 traces our new system. The condensate returns to the vacuum tank. The vacuum pump discharges the condensate to a condensate return tank. A feed water pump sucks water from the condensate return tank and delivers it under pressure to the boiler. The feed water regulator turns the feed water pump on and off.

The feed water pump has no way of knowing how much water the boiler needs. There must be something to turn it on to start pumping and to turn it off. And that is exactly what a feed water regulator does.

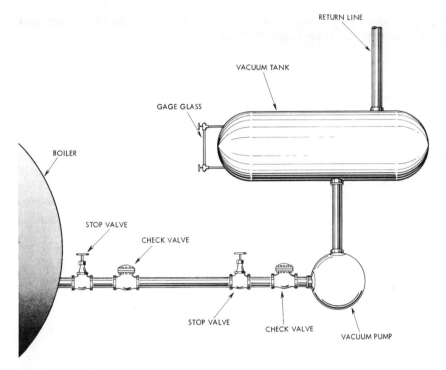

Fig. 3-8. Schematic showing returns going to vacuum tank to vacuum pump to boiler.

Location. The feed water regulator is located at the normal operating water level. It is connected to the boiler in the same manner as the water column and the low water cut-off. The top line is connected to the highest part of the steam side of the boiler, the bottom line to the water side of the boiler, well below the normal operating water level.

How it works. When the water level drops in the boiler, the feed water regulator starts the feed water pump. When the water in the boiler is at its normal operating level, it stops the pump. Fig. 3–9 shows how it does this.

To understand this fully, we should go back and look at the condensate return tank. The heating returns (condensate) are delivered to the return tank by the vacuum pump. If for some

Fig. 3-9. Another feed water system. Condensate returns to vacuum tank then to condensate return tank to boiler.

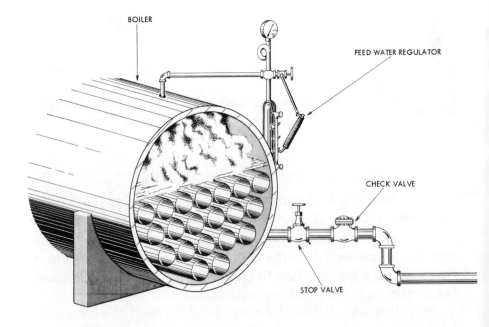

reason the vacuum pump can't return this condensate and the boiler calls for water, the feed water pump will start. With no returns in the return tank, the feed water pump would run dry. The water level in the boiler would drop and if the low water

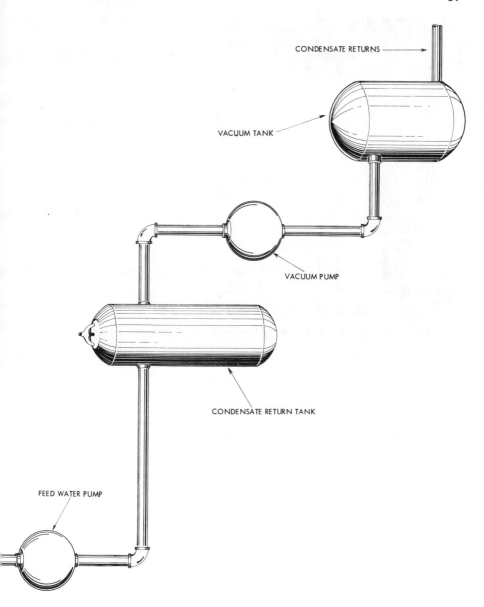

CONDENSATE RETURNS

VACUUM TANK

VACUUM PUMP

CONDENSATE RETURN TANK

FEED WATER PUMP

cut-off works, the boiler would shut off on low water. If not, the boiler could become damaged or explode.

Obviously, the pump and boiler must be protected from such an explosion. To find how this is done, look at the return

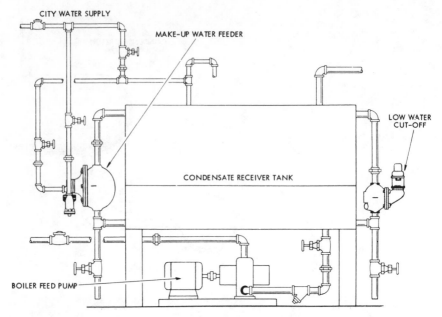

Fig. 3-10. Condensate return tank showing make-up water feeder.

tank in Fig. 3–10. The tank is equipped with an automatic city water make-up valve. This keeps a constant water level in the return tank at all times. When the boiler calls for water, the feed water pump is protected, and the boiler will get the water it needs.

Looking Back

1. Low water cut-offs shut off the burner in the event of low water.
2. The low water cut-off is located a little below the normal operating water level.
3. Each low water cut-off has a blow down valve. This valve should be opened daily to test the low water cut-off while the boiler is firing.

4. Once a month it should be tested by dropping the water level in the boiler.

5. If the boiler is equipped with two low water cut-offs, they must both be tested and given the same attention, and both must be in good operating condition.

6. The feed water regulator is located at the normal operating water level and is connected to the boiler in the same manner as the water column.

7. It maintains a constant water level in the boiler by starting and stopping the feed water pump.

8. The feed water pump gets its water from the condensate return tank which gets its water from the vacuum pump.

9. The condensate return tank has an automatic city water make-up valve to keep a constant water level in the return tank.

4

STEAM ACCESSORIES

Before we discuss steam accessories, let us review the basic water-to-steam-to-water cycle. You will remember that steam leaves the boiler through the main steam line, which connects to the *header*. The *main branch lines* extend from the header and merge with *risers*, which take the steam wherever it is needed to heat a room.

MAIN STEAM STOP VALVE

Steam travels from the boiler through the radiators. To cut the boiler in on the line (that is, let steam flow from the boiler into the header), or take a boiler off the line, there must be a *main steam stop valve* on the main steam line. This valve should be a *gate valve*. A gate valve works exactly like the overhead door of a garage. To get the car out of the garage, we lift the door. When we put the car back in the garage, we pull the door down. So, in a gate valve, to let steam pass through it the gate (a metal plate) is lifted. To shut off the flow of steam, the gate goes down, closing the valve.

Fig. 4–1 (left) shows a cut-away sketch of a gate valve. When this valve is open, the steam flows through it, and nothing restricts its flow. This is very important. If a globe valve (Fig. 4–1, right) were used, the steam would have to come in under the seat and then pass up. This restriction of flow could cause a drop in steam pressure.

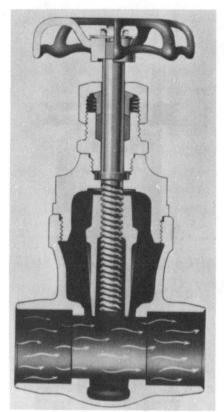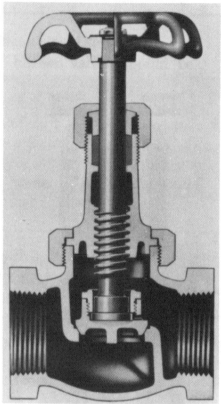

Fig. 4-1. A gate valve (left) is used on the main steam line. A gate valve does not restrict the flow of steam. The globe valve (right) will slow down the flow. Kennedy Valve Co.

It should be clear now why the main steam stop valve should be a gate valve. With the gate valve open, there would be no restriction to the flow of steam, for a boiler stop valve is always either fully open or completely closed.

It is also important for the fireman to know exactly when the main stop valve and the valves on the header are open or closed. When the valve is 20 or 30 feet in the air (as it often is on large boilers) it is hard to tell without actually trying it.

To enable the fireman to tell at a glance whether the main stop valve and the valves on the header are closed, they should

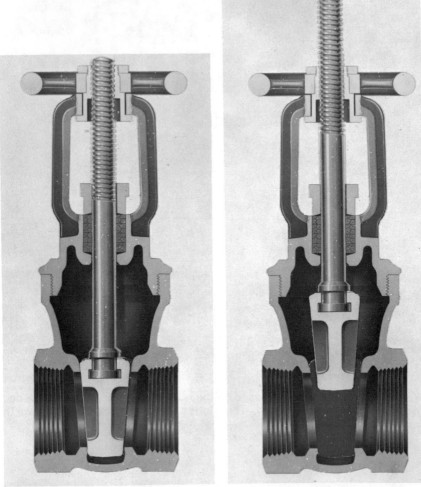

Fig. 4-2. Outside stem and yoke valve, a gate type with rising stem, open (right) and closed (left). Kennedy Valve Co.

all be O.S. & Y. (outside stem and yoke) valves. Such a valve shows by the position of its stem whether it is open or closed. In Fig. 4–2, the valve on the left is closed; the one on the right is open. The photograph shows why these valves are sometimes called *rising stems*.

Looking Back

1. All boilers must have a main steam stop valve.

2. Main stop valves and header valves should be O.S. & Y. gate valves.

3. An O.S. & Y. gate valve shows the fireman by the position of its stem whether it is open or closed. When open, it offers no resistance to the flow of steam.

4. Main stop valves and header valves are always either wide open or completely closed.

STEAM TRAPS AND STRAINERS

Purpose. A steam trap is an automatic device that will increase the overall efficiency of a plant by removing air and water from the steam lines without the loss of steam.

As steam travels from the boiler to the main header it drops a little in temperature. This causes some condensate to form. The condensate must be removed for a number of reasons. If it were picked up with the steam as it travels through the main branch lines, it could lead to water hammer and a possible pipe rupture.

Water hammer is caused by steam pushing condensate ahead of it as it travels through the steam lines. When it reaches a 90° turn, the restriction in the pipe and the sudden change in direction of the steam-condensate mixture results in a loud hard knock. To prevent water hammer or a pipe rupture the condensate must be removed from the steam lines. This is accomplished by placing steam traps where condensate might build up.

Location. Steam traps should be located:

1. On the ends of the main header.

2. On the end of each main branch line.

3. On each radiator or heat exchanger, wherever steam gives up its heat.

Types. There is only one type of trap that is being used now. That is the non-return type which is made in several models. Another type of trap may occasionally be found on some old systems. This is the return trap. It is a large trap placed alongside the boiler and a little above it. It discharges the condensate directly into the boiler when the pressure in the trap is equal to or slightly higher than the boiler pressure.

The non-return type of trap is the one mentioned in Chapters 1 and 3. This type sends the condensate through a vacuum pump to a condensate receiver tank. From the tank it is pumped to the boiler. Three kinds of non-return traps are used. These are shown in Fig. 4–3.

The inverted bucket trap is shown at the top of Fig. 4–3. Steam enters from the bottom and into the inverted bucket. This holds it up. As the condensate fills up the trap the bucket loses its buoyancy and sinks. This opens the discharge valve and the vacuum on one side and the steam pressure on the other combine to remove the condensate. When the condensate has been removed and steam enters the trap again the bucket is lifted and closes the discharge valve. A small hole on top of the bucket allows any trapped air to get out. Otherwise the air would prevent the trap from sinking and opening the discharge valve.

A *thermostatic* trap is shown in the middle of Fig. 4–3. This is the most common trap and is used on many kinds of radiators. The trap contains a flexible bellows which has a fluid in it that boils at steam temperature. When the fluid boils the vapors cause the bellows to expand and push the valve closed. When the temperature in the trap falls below that of steam the fluid stops boiling and condenses. This allows the bellows to contract and pulls the valve into the open position. The bellows keeps expanding and contracting depending on whether it is surrounded by steam or condensate. This keeps closing and opening the valve so that only the condensate leaves the radiator. The vacuum on the return line and the pressure from the steam in the radiator removes the condensate from the trap when the discharge valve is open.

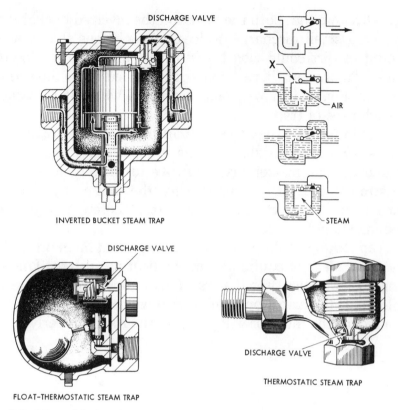

DISCHARGE VALVE

INVERTED BUCKET STEAM TRAP

X

AIR

STEAM

DISCHARGE VALVE

FLOAT-THERMOSTATIC STEAM TRAP

DISCHARGE VALVE

THERMOSTATIC STEAM TRAP

Fig. 4-3. Three kinds of steam traps—inverted bucket, float thermostatic, and thermostatic.

The *float thermostatic* trap is shown at the bottom of Fig. 4–3. In this trap a float opens and closes the discharge valve according to the amount of condensate in the bowl of the trap. The condensate causes the float to rise and open the discharge valve. The suction and the pressure remove the condensate and the float drops and closes the valve. Air in the trap which might interfere with its operation is removed by a thermostatic valve at the top.

We have now seen how an inverted bucket, a float, and a thermostatic trap work. But to keep working, the traps must be kept free from dirt and other impurities. Another look at

Fig. 4–3 shows that both the float and the inverted bucket traps have very small discharge orifices. It is obvious that a slight amount of dirt could clog these orifices and stop up the discharge. To prevent this a *strainer* is installed in the line before the trap. Fig. 4–4 shows the strainer and where it is installed for each type of trap.

It is important that the strainer be cleaned at regular intervals. In a new installation this is especially important because of foreign matter that might be in the pipe lines. Clean the strainer every three months for the first year, and then twice during the next year. After that, a thorough once-a-year cleaning should suffice.

Trap troubles. Traps are usually neglected more often than any other piece of equipment in the heating system. This is a dangerous practice. For if a trap fails to open, condensate will build up. If the trap is on a radiator, it will become waterlogged. As a result, the radiator will not heat the room properly.

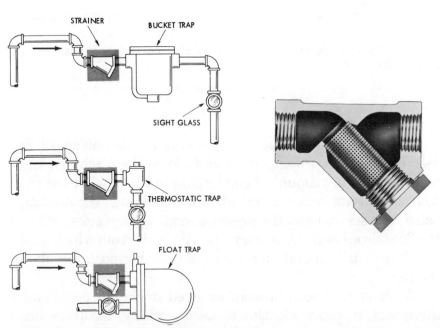

Fig. 4-4. Strainer installed before trap prevents lines from getting clogged.

On the other hand, if the trap sticks open, steam will blow through and into the vacuum pump or condensate receiver. Here, a number of things will happen. First, the temperature in the room will rise because the steam cannot be properly controlled.

Second, the live steam returning to the vacuum tank or receiver will raise the temperature of the water. This could make the pump *steam-bound* (water will flash into steam). This would have the effect of transforming the condensate in the return lines into steam. This, of course, could not be pumped back into the boiler. So the water level in the boiler would drop, and the boiler would shut off because of low water. If, however, the boiler were equipped with an automatic city water make-up valve (sometimes called a *water feeder*), this would open and let city water into the boiler. After the steam-bound condition had been corrected, the fireman would find his system had too much water, and he would have to remove some using the bottom blow down valve.

Finally, if steam blows through the trap, fuel is being wasted. Steam should stay only in the radiators and steam lines. The presence of steam in the water return lines reduces boiler efficiency.

Testing traps. If he checks the steam traps as soon as there is a sign of a room overheating or condensate return temperatures increasing the fireman will not only save money on fuel but he will also have a more comfortable building.

Steam traps can be checked very easily. One way of checking them is with strap-on thermometers. These thermometers clip over the steam line, and will record the temperature in the line. Take a temperature reading before the trap and a temperature reading after the trap, as shown in Fig. 4–5. There should be a 10°F to 20°F difference between the two readings. This is because, at location (1) in Fig. 4–5, steam is entering the trap; at location (2), condensate should be leaving the trap. If steam were blowing through the trap, the temperature would be the same on both sides of the trap.

There is another method of checking the traps in a heating

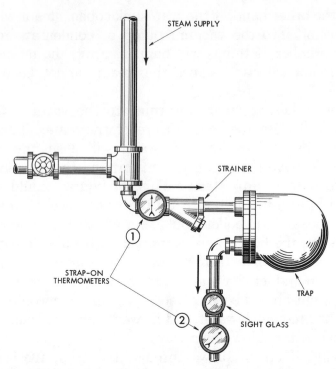

STEAM SUPPLY

STRAINER

STRAP-ON
THERMOMETERS

TRAP

SIGHT GLASS

Fig. 4-5. Test traps by strapping one thermometer on steam line before the trap and another one on the condensate line where it leaves the trap. There should be a 10° difference if the trap is working.

Table 4-1. Relation of steam temperature to steam pressure.

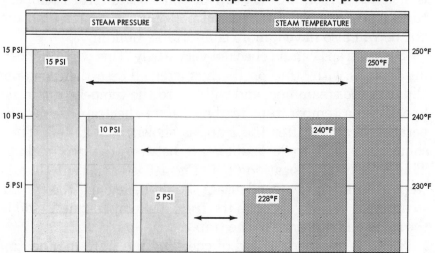

STEAM PRESSURE	STEAM TEMPERATURE
15 PSI	250°F
10 PSI	240°F
5 PSI	228°F

system. That is to use a temperature stick. This is a special crayon that has a controlled melting point. Obtain one that suits the temperature of the steam in your system. Steam has a corresponding temperature and pressure. In other words, as pressure increases, so does temperature. Table 4–1 shows how steam temperature corresponds to steam pressure.

After selecting the proper temperature stick, place a crayon mark on the discharge side of the trap. If the trap sticks open and steam blows through, the return line temperature goes up melting the crayon on the pipe. This shows that the trap is not working properly.

Trap maintenance is really very simple. There is no excuse for neglecting it.

Trap selection and sizing. It is important to select the proper type and size of trap for a particular system. There are many types of traps, as we have mentioned before. But it is unnecessary to discuss the selection and sizing of traps in this book. If the fireman has any trap problems, he should contact the manufacturer. Field engineers will come to the plant and help the fireman solve his trap problems. Manufacturers do not charge for this service. They sell a good product and want it to provide the best service.

Looking Back ────────────────────────────────

1. There are only two types of steam traps: return and non-return.
2. The return trap returns the condensate directly to the boiler. It is outdated and seen only on old installations.
3. The non-return trap discharges either to a vacuum pump or to a condensate return tank. There are three kinds of non-return traps: the inverted bucket; the float thermostatic; and the thermostatic.
4. A trap increases the overall efficiency of a plant by

removing air and water from the steam lines without the loss of steam.

5. Strainers should be located in the steam line in front of the inverted bucket or float traps.

6. Strainers in new plants should be cleaned often during the first year of operation. This is because there are many impurities in the lines of new installations.

7. Traps should be tested frequently with strap-on thermometers or a temperature stick.

COMBUSTION ACCESSORIES

FUEL OIL SYSTEMS

Basic fuel oil systems were covered in the introduction rather briefly. We will now cover them in detail.

Number six fuel oil is an industrial type fuel which must be preheated to pump and further heated to burn. A fireman that can handle number six oil can handle other types quite easily. Therefore this is the one we will discuss. But first there are a few terms that need defining.

1. Heating value of fuel is measured by Btu's. For example, No. 6 fuel oil has a Btu content of 148,000 to 152,000 per gallon.

2. Btu means British thermal unit. This is the amount of heat needed to raise the temperature of one pound of water one degree Fahrenheit.

3. Flash point is the temperature to which oil must be heated to cause a flash when an open flame is passed over it. It will not burn continuously but only flash.

4. Fire point is the temperature to which fuel oil must be heated for it to burn continuously when exposed to an open flame. The fire point is higher than the flash point temperature.

In handling fuel oil it is important to understand thoroughly the flash point and the fire point of oil. It should make you aware of the dangers of overheating the oil in the storage tank. The temperature of the oil in the storage tank should be about

100°F-120°F. This is warm enough to pump yet not hot enough to give off a vapor that might flash. The tank temperature is controlled by steam going to either a heating coil or a heating bell in the tank.

OIL SYSTEM LAYOUT

Fig. 5–1 shows a typical fuel oil system.

The line coming from the tank to the pump is called the

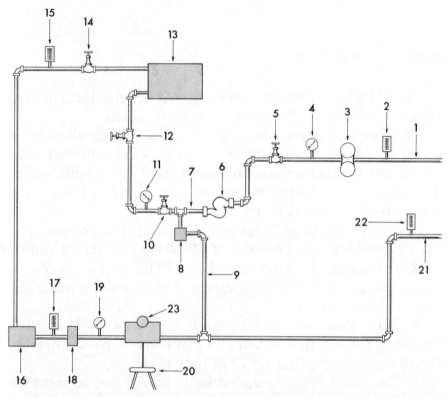

Fig. 5-1. Layout of a typical fuel oil system. 1. Suction line; 2. thermometer; 3. duplex strainer; 4. suction gage; 5. suction valve; 6. fuel oil pump; 7. discharge line; 8. relief valve; 9. relief valve discharge line; 10. discharge valve; 11. pressure gage; 12. inlet valve; 13. steam heater; 14. outlet valve; 15. thermometer; 16. electric heater; 17. thermometer; 18. strainer; 19. pressure gage; 20. fuel oil burner; 21. oil return line; 22. thermometer; 23. constant pressure regulator.

suction line (1). On this line you will have a thermometer (2) which records the temperature of the oil leaving the tank. The duplex strainer (3) permits one strainer to be cleaned while the other is in service. This means you do not have to shut off your boiler. The suction gage (4) shows how many inches of vacuum the fuel oil pump is pulling. The suction valve (5) located just before the pump allows you to isolate the pump from the tank. If there is a leak on the suction line between the tank and the suction side of the pump, air will be brought into the system. The suction gage will then pulsate (move back and forth) and the fires in the boiler will also pulsate. This air must be bled off (removed from the lines) or it might cause the fire to go out. The oil enters the fuel oil pump (6) and leaves under pressure through the discharge line (7). The relief valve (8) protects the pump in the event of high pressure. The relief valve discharge (9) goes back to the return line to the fuel oil tank. A discharge valve (10) allows the pump to be isolated from the rest of the system for pump packing and repair. A pressure gage (11) shows how much pressure the pump is developing. The oil now travels to a steam heater (13). Here the temperature is brought up closer to the flash point. The inlet valve (12) and outlet valve (14) can isolate the heater (13) for cleaning or repairs. The thermometer (15) on the outlet side of the heater shows the temperature of the oil leaving the steam heater. The oil then goes to the electric heater (16). Here the temperature of the oil is brought up to the temperature at which it will burn. This temperature is 150°-180°F for rotary cup burners and 180°-200°F for mechanical atomizing burners. These burners will be described later.

After the electric heater a thermometer (17) shows the temperature of the oil leaving the heater. Another strainer (18) catches any impurities that might still be in the lines. The pressure gage (19), before the burner, allows regulation of the pressure at the burner (20). Here the oil is introduced into the firebox to burn. However, not all the oil is burned. Some of the oil enters a return line (21) and is pumped back to the tank. The thermometer (22) shows the return oil temperature. The con-

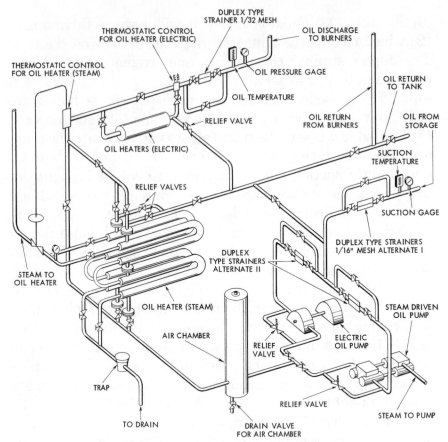

Fig. 5-2. Piping layout for a fuel oil system.

stant pressure regulator (23) maintains the required oil pressure on the line. You must trace your own system and become thoroughly familiar with each valve, thermometer, strainer, and other parts. Then, in the event of a breakdown, you will know how to change over or bypass the equipment causing trouble. Fig. 5–2 illustrates another fuel oil system.

Looking Back

1. It is important to keep the oil in the storage tank below its flash or fire point.

2. Any leaks in the fuel oil line between the tank and the suction side of the pump will allow air to get into the system. This will cause the fires to fluctuate.

3. The fireman should trace all oil lines and become familiar with all valves in order to change over or bypass any faulty equipment.

4. Burn #6 oil between 150° and 180° F in rotary cup burners and between 180° and 200° F in mechanical atomizing burners.

5. Strainers should be cleaned daily.

FUEL OIL HEATERS

The only oil that must be preheated before it will burn is Bunker C or #6 oil. It has one of the highest Btu contents of any of the oils. It is also cheaper. However, it does require heaters in order to burn it. To protect your plant you should have both steam and electric heaters.

Fig. 5–3 shows a typical shell and tube steam heater.

The temperature of the oil is regulated by a temperature regulating valve. Fig. 5–4 explains how it works. As the temperature increases the bulb (1) picks up or senses the temperature increase. The gas in the tube (2) expands and causes pressure on the diaphragm (3). This pressure overcomes the spring pressure (4) and throttles (cuts down) the flow of steam to the heater through the valve (5). As the temperature of the oil cools the bulb (1) feels or senses the temperature drop. The gas in the tube cools (2), the pressure on the diaphragm (3) drops and the spring (4) opens the valve (5) and allows more steam to enter the heater. The stop valves (6), one on each side of the control valve (5), enable the valve to be repaired. The steam to the heater would then be regulated by hand, using the by-pass valve (7). This is a rather simple regulator but it is very effective.

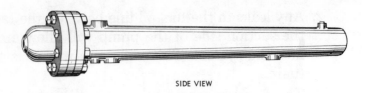

SIDE VIEW

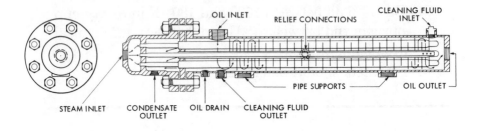

SECTIONAL VIEW

Fig. 5-3. Tubular type steam fuel oil heater, outside and inside.

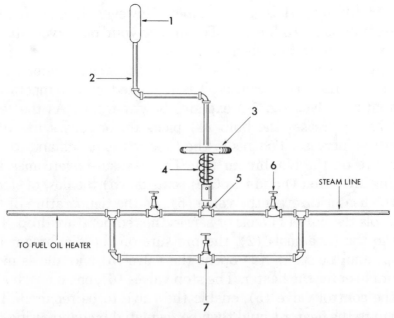

Fig. 5-4. Diagram of fuel oil temperature regulator. 1. Sensing bulb; 2. tube; 3. diaphragm; 4. pressure spring; 5. control valve; 6. stop valve; 7. by-pass valve.

The electric heater located near the burner is used as a booster to raise the oil temperature to the proper burning temperature. The steam heater is not used to bring the oil temperature up to the igniting point. The electric heater is also needed to start the boiler up when there is no steam. The oil pumps are started and the oil circulated through the electric heater and back to the tank. When the oil has been warmed enough, the boiler can be lit off (fired) and the steam cycle started.

The temperature of the electric heater is controlled by a thermostat. When the proper oil temperature is reached a switch opens breaking the circuit to the heater. As the temperature drops the switch pulls in and closes the circuit to the heater. A temperature control on top of the heater can be adjusted to raise or lower the oil temperature.

FUEL OIL STRAINERS

Duplex strainers should be located on the suction line before the pump as shown in Fig. 5-1 and 5-2. These strainers, Fig. 5-5, have a fine mesh screen basket inside. It is possible to run your boilers while cleaning the strainer since only one strainer is cleaned at a time. It is important that the strainers be kept clean and that there is always a clean strainer available.

An increase in the suction gage reading indicates either a dirty strainer or cold oil. Be sure to check to see which it is and correct the situation. If you are using #6 oil the strainers should be cleaned every 24 hours. It is important that the flange cover Fig. 5-6 (1), be cleaned whenever the strainers are cleaned, and that the gasket (2) be carefully replaced. If you fail to do this air may get into the system.

The strainer baskets (3) have different sizes of mesh screening. When ordering replacement baskets be sure to order the proper size screening. If you have any questions about the duplex strainers in your plant contact the manufacturer. He has trained field engineers who will help solve your problems. Secondhand information can be inaccurate and costly.

Fig. 5-5. A duplex strainer for fuel oil. Kraissl Co.

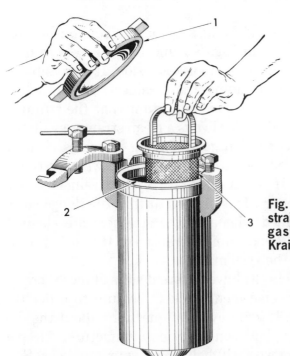

Fig. 5-6. Cleaning a fuel oil strainer. 1. Flange cover; 2. gasket; 3. strainer basket. Kraissl Co.

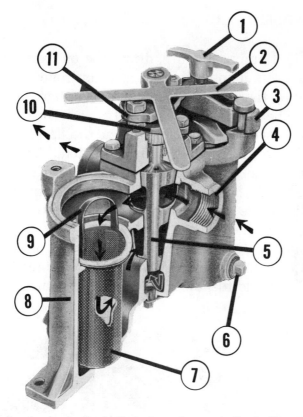

Fig. 5-7. Parts of a duplex fuel oil strainer. Arrows indicate flow of oil. 1. Hand screw; 2. handle; 3. yoke; 4. oil inlet; 5. plug valve; 6. drain; 7. strainer basket; 8. body of strainer; 9. spring handle; 10. stuffing box gland; 11. locking flange. Kraissl Co.

The parts of a duplex strainer shown in Fig. 5–7, are: 1. hand screw; 2. handle; 3. yoke; 4. oil inlet; 5. plug valve; 6. drain (one on each basket); 7. double element basket; 8. body; 9. spring handle of basket (holds basket on seat); 10. stuffing box gland; and 11. locking flange (adjustable to provide valve clearance).

Arrows indicate the flow of the oil. Oil comes through inlet (4) into top of basket (7) through the sides of the basket and through the tapered plug valve (5) and on out through the outlet opening.

FUEL OIL PUMP

The fuel oil pump, Fig. 5–8, draws the oil from the fuel oil tank and delivers it to the burner at a controlled pressure. As Fig. 5–1 shows, there is a suction gage, a discharge gage, and a relief valve. The relief valve is located between the fuel oil pump and the discharge valve. The suction and discharge valves must be opened before the fuel oil pump is started. The relief valve will protect the fuel oil lines and the pump if the discharge valve is left closed or if there is some obstruction in the lines.

The relief valve will discharge into the return line which returns the oil to the fuel oil tank.

If the suction gage records a high vacuum of 10″ or more (remember, vacuum is recorded in inches), it would be an indication of either cold fuel oil or dirty suction strainers.

Fig. 5-8. A fuel oil pump. Kraissl Co.

Fig. 5-9. Parts of a fuel oil pump. 1. Internal gear; 2. idler gear; 3. bearing and idler assembly; 4. stuffing box; 5. return seal; 6. coupling; 7. ball bearing; 8. support for pump; 9. V-belt drive; 10. drive pulley; 11. belt guard; 12. bed plate. Kraissl Co.

The parts of a fuel oil pump are shown in Fig 5–9. They are 1, Internal gear; 2, idler; 3, integral bearing and idler assembly; 4, stuffing box with return seal and soft packing; 5, interchangeable return seal which allows change of rotation in the field; 6, flexible coupling; 7, ball bearing unit; 8, bearing support; 9, V-belt drive; 10, pulley; 11, cast iron belt guard; 12, cast iron bed plate.

The arrows in Fig. 5–10 indicate the flow of oil when the pump is turning in either a counter-clockwise or clockwise direction.

A reversible pump can be mounted in the best location and the rotation adjusted if necessary.

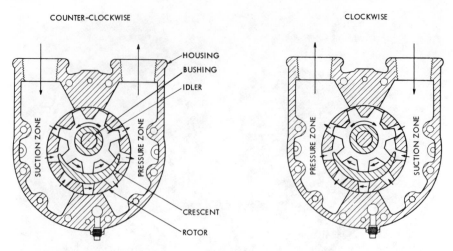

Fig. 5-10. Fuel oil pump. Arrows show flow of oil according to pump rotation.

FUEL OIL BURNERS

In low pressure plants using #4, #5, or #6 fuel oil the *rotary cup oil burner,* Fig. 5–11, is one type that is commonly used. The oil is delivered through a fuel tube at a low pressure and discharges to the inside of a rapidly rotating brass cup. As the cup rotates centrifugal force disperses the oil in small droplets. A fan forces air around the rotating cup and spins it in the

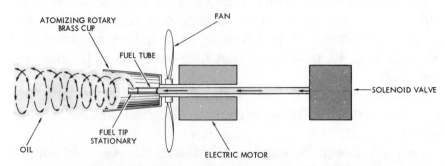

Fig. 5-11. A direct drive rotary cup oil burner.

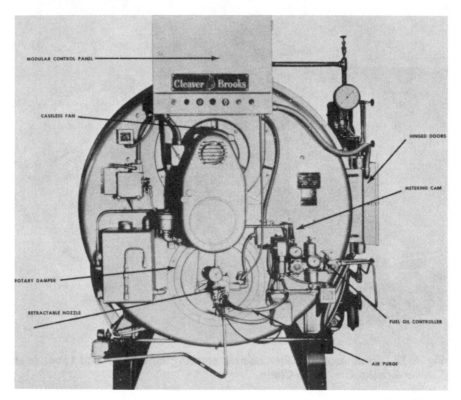

Fig. 5-12. An oil fired boiler, front view. Cleaver-Brooks Div. Aqua-Chem.

opposite direction to the oil. This causes the fuel and air to mix for proper combustion. Fig. 5–11.

The oil pressure may vary from 5 to 10 lbs and the oil temperature is kept between 150° and 180°F.

The *air atomizing burner* is used on both high and low pressure boilers. It has proven to be an effective burner with excellent combustion efficiency. The Cleaver Brooks boiler, Fig. 5–12, is equipped with an air atomizing burner. The burner is integrally built into the front head of the boiler and swings out when the head is opened for inspection or maintenance. Fig. 5–13. It is a low pressure atomizing burner.

The operational feature of this type of burner is its ability

Fig. 5-13. Front door of boiler opened showing inner door and tube sheet. Cleaver-Brooks Div. Aqua-Chem.

to produce a high degree of atomization of both light and heavy oil. This provides a close contact of air and oil which results in clean, complete combustion before the flames reach the furnace wall making it possible to keep secondary air to a minimum. Fig. 5–14 shows how the primary or atomizing air passes around the fuel tube and mixes with the oil in the nozzle. With heavy oil the burner nozzle and piping are purged, Fig. 5–15, of oil so the heated oil can flow quickly for the next start-up cycle. This eliminates coking of oil in the nozzle and congealing of the oil in the piping. Controls are provided to regulate both oil viscosity and pressure to the oil metering valve, thus delivering the exact amount of oil to match the boiler load demand. The nozzle can be completely removed from the firing position for cleaning or maintenance without removing the front head. In

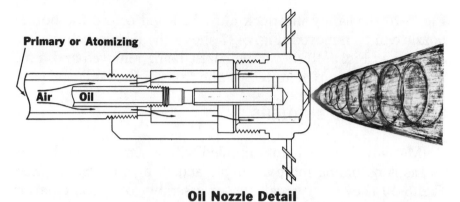

Oil Nozzle Detail

Fig. 5-14. Diagram of atomizing oil nozzle. Cleaver-Brooks Div. Aqua-Chem.

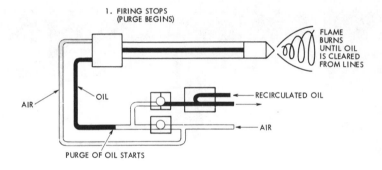

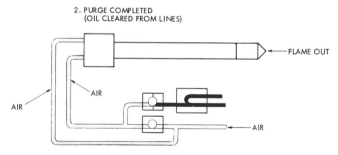

Fig. 5-15. Diagram showing purging of oil lines after firing. Cleaver-Brooks Div. Aqua-Chem.

Fig. 5–16 the safety interlock must be lifted before the burner nozzle can be removed. Fig. 5–17 shows the burner being withdrawn and Fig. 5–18 shows the nozzle being removed and hung in a vise jaw clamp mounted on the front of the boiler.

COMBINATION BURNERS

Many boilers are being supplied with a combination burner as gas is becoming more available and at a competitive price. Fig. 5–19 shows a typical burner assembly of a combination unit. The combination boiler permits the operator to switch from one fuel to another for economy, for a failure of the fuel in use, or a shortage of a fuel. The Cleaver Brooks boiler shown

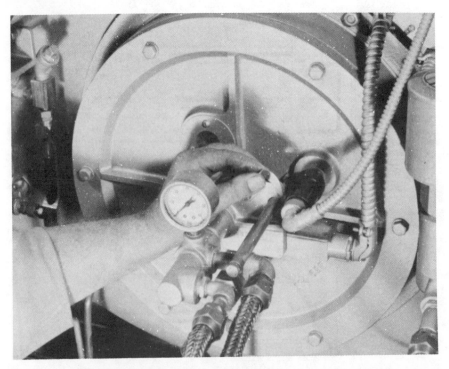

Fig. 5-16. Opening safety interlock on oil burner. Cleaver-Brooks Div. Aqua-Chem.

Fig. 5-17. Burner nozzle being removed from boiler. Cleaver-Brooks Div. Aqua-Chem.

in Fig. 5–19 has an integral gas burner built in so you can convert quickly from one fuel to another in less than a minute. This quick change over feature is important in an emergency; the boiler can be switched over without losing the plant load. Fig. 5–20 shows both cams needed to control the air-fuel ratio of a combination burner and a close-up of the adjustable contour cam is illustrated in Fig. 5–21.

GAS BURNERS

There are some boilers equipped to burn only gas. Fig. 5–22. Fig. 5–23 shows how gas and air mix in a high velocity gas

Fig. 5-18. Burner nozzle placed in clamp for disassembly and cleaning. Cleaver-Brooks Div. Aqua-Chem.

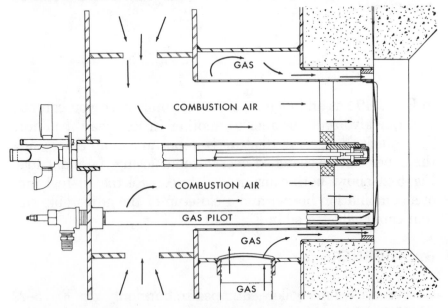

Fig. 5-19. Diagram of a combination burner. Gas is being used.

Fig. 5-20. Boiler front of combination boiler. Note cams that control air-fuel ratio. Cleaver-Brooks Div. Aqua-Chem.

burner. Gas burning boilers are not used in some parts of the country, however, a typical gas burner other than the high velocity type is shown in Fig. 5-24.

The gas line (1) is fitted with a gas cock (2) which allows the fireman to close (cut off) the gas to the system when doing any repairs. Next on the gas line is a solenoid valve (electric valve) (3) that controls the gas to the pilot (4). The manual reset valve (5) is an electric valve that cannot be opened until the gas pilot is lighted. The balanced zero reducing governor (6) reduces the city gas pressure to zero pressure. The small line just before the governor goes to the vaporstat (7). The vaporstat is a switch that is turned on by the gas pressure in

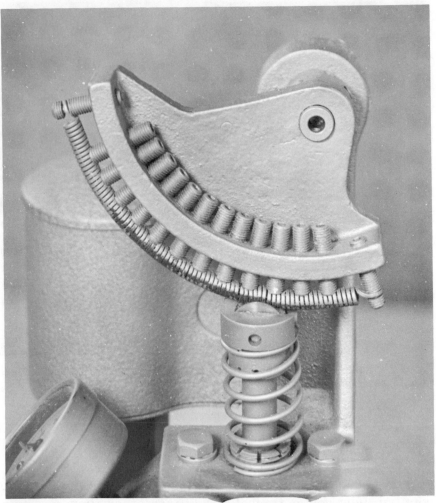

Fig. 5-21. Close-up of adjustable air-fuel ratio cam. Cleaver-Brooks Div. Aqua-Chem.

the line or turned off when there is no pressure. Its function will be explained as other parts of the system are described. The main gas solenoid valve (8) will open at the proper time allowing gas to be drawn down to the mixjector (9). The forced draft blower (10) will send air through the motorized control valve (11). The air passes through a venturi (12) and draws the gas

Fig. 5-22. A gas fired boiler. Cleaver-Brooks Div. Aqua-Chem.

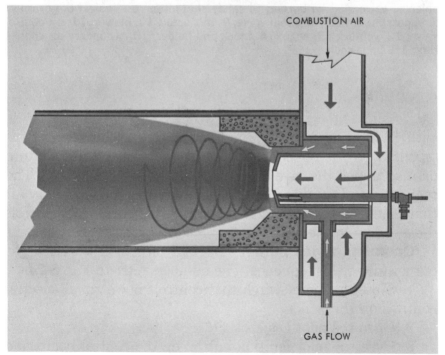

Fig. 5-23. High velocity gas burner. Cleaver-Brooks Div. Aqua-Chem.

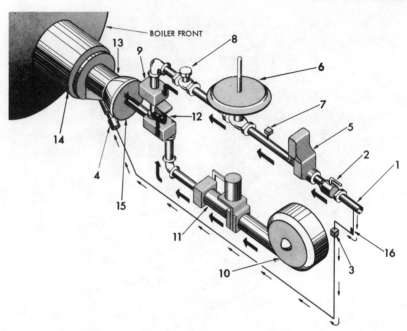

Fig. 5-24. Diagram of a gas fired system. 1. Gas line; 2. gas cocks; 3. pilot solenoid valve; 4. pilot burner; 5. safety shut-off or reset valve; 6. governor; 7. vaporstat; 8. main solenoid valve; 9. mixjector; 10. blower; 11. air control valve; 12. venturi; 13. cage; 14. block and holder; 15. secondary air control ring; 16. pilot cock.

with it into the cage (13). The block and holder (14) is mounted on the boiler front and as the gas and air mixture passes through the cage it is ignited by the pilot light (4). The cage has an adjustable ring (15) that controls the secondary air that enters to complete combustion. A gas cock (16) controls the gas flow to the pilot.

Operating a gas burner. To light off a boiler with a gas burner follow this sequence. The numbers refer to Fig. 5–24.

1. Close the main switch that controls power to all electric controls on the boiler.

2. Open the pilot gas cock (16).

3. Push the pilot ignition button on the control board and the pilot will light.

4. Open the manual main gas cock (2).

5. Lift the handle on the manual reset safety shut-off valve. (5).

The gas now flows up to the balanced zero reducing governor (6) where it is reduced to zero pressure as it passes through the governor. The gas pressure switch or vaporstat (7) will complete an electric circuit as soon as it "proves" gas pressure to the governor. The circuit then starts the forced draft fan or blower (10).

The vaporstat "proves" gas pressure by means of a diaphragm and switch. As the pressure builds up in the gas line it pushes against a large diaphragm in the vaporstat. As the diaphragm expands it causes a mercury tube to tilt which completes an electric circuit to the blower.

The motorized air control valve (11) is a slow moving electric motor that operates a flapper type valve on the air line to control the amount of air that can pass through. It is in the low fire position that allows only a minimum of air to flow through. This means that only a small amount of air under pressure can pass through to the venturi (12). As the air passes through the venturi it pulls in gas through the mixjector (8) which mixes the air and gas and passes the mixture to the cage (13) where it is ignited by the pilot (4) and passes through the block and holder (14) into the boiler firebox. Secondary air can be adjusted by opening or closing the adjustable ring (15) at the front of the cage.

Once the burner has ignited and the boiler warmed up the burner may be placed in high fire by putting the selector switch to automatic and pushing the run button on the panel board. As soon as this is done the motorized air control valve will start to open the flapper valve allowing more air to flow through. The more air that flows the more gas will be pulled thro. gh the mixjector and the boiler will be in high fire.

The burner will now be under the control of the modulating pressuretrol and the on and off pressuretrol, both of which have been described in previous chapters.

It should be noted that the pilot must ignite before the

manual reset safety shut off valve can be opened to allow gas through. A flame rod must prove the pilot, thus completing an electric circuit allowing the manual reset safety shut off valve to engage. In the event of a pilot or main flame failure the valve will close shutting the gas off, putting the boiler off on safety. In the event of low water the manual reset safety shut off will also close shutting gas off to the burner.

There is another safety precaution. The forced draft blower cannot start until the gas pressure switch (vaporstat) located in the line before the balanced zero reducing governor proves gas pressure in the line up to the governor.

Looking Back

1. Number 6 fuel oil must be heated to the proper temperature in order to burn.
2. Fuel oil heaters are either steam or electric. Both are needed as the electric must be used to start the boiler before steam is available.
3. Oil strainers must be cleaned and changed every 24 hours.
4. Fuel oil strainers have two strainer baskets so that one can be cleaned while the other is operating so the flow of oil is not stopped.
5. Fuel oil pumps have suction gages, pressure gages, and relief valves. Check these for proper operation of the pump.
6. Some oil burners have a rotating cup that throws the oil into the air as tiny drops.
7. Atomizing oil burners mix the oil with air inside the nozzle.
8. Some boilers can be used for either oil or gas by making some adjustments. In some cases separate burners are used while in others the changes are made in the burner.

COAL

Coal as a fuel for steam boilers should not be ignored. Many large industrial plants still use coal most effectively. There are also low pressure boilers that burn coal. Since we are primarily interested in low pressure boilers we will only talk about hand firing and stoker firing. Pulverizers are found in the larger high pressure plants. Pulverized coal is coal that is ground to the consistency of face powder and then blown into the boiler where it is burned in suspension the same as gas or oil. There are no grates necessary to support the fuel bed. Before we go too far, let's discuss the coal that will be burned.

Two types of coal. They are *hard coal*—called Anthracite coal and *soft coal*—called Bituminous coal. Hard coal has a high fixed carbon content and burns mostly on the grates. (Grates are used to support the coal bed and also to allow air to pass through to aid in combustion.) Soft coal has a high volatile content (gas content) and this burns above the grate.

A boiler that uses soft coal as a fuel must, therefore, have a larger furnace volume than a boiler burning hard coal. There must be more distance above the grates. This is necessary so that the gases can burn completely before they hit the heating surface. If these gases have not burned completely before they reach the heating surface they will cool and result in soot forming and causing smoke. Smoke is, therefore, a sign of incomplete combustion. Hard coal with a high fixed carbon content does not offer this problem.

HAND FIRING COAL

Firing coal by hand is almost a lost art. There are not many plants that still use hand fired boilers. The reason is quite obvious. The size of the boiler must be limited. A fireman can only spread the coal so far. Notice the use of the word *spread*. Coal is never dumped in a boiler, it is spread over the grates. A good fireman can fan the coal and have it actually spray out like a fan over the surface of the fire. This is important as it helps to form an even fire bed. No written words can describe to a fireman how to hand fire coal. My own experience as a fireman proved that this is a skill that can only be developed through actual experience. We will go over a few points that might help.

Regardless of whether you are firing hard coal or soft coal when you first open the firing door to coal over, throw the first shovel of coal on the front of the fire. This will serve two purposes. It will cover the hot coals and cut down on the heat that would otherwise blast out at the fireman. It will also cut down a bit on the intensity of the fire making it easier for the fireman to see into the fire box and evaluate the condition of the fire.

METHODS OF COALING OVER

Coal may be fired alternately on one side of the boiler firebox at a time. Or it may be fired by coaling over the whole firebox. I personally prefer to fire alternately. There is less tendency to cause smoke because there is always one side of the firebox burning to help burn off the volatile gases. I also found it better to coal over often but lightly. I then had better control of the fires as well as the steam pressure. If both sides of the firebox are coaled over at the same time and the fireman puts too much coal on there could be trouble. First the coal acts as a dampening agent, it cools the firebed. This could cause a drop in steam pressure. If a high volatile coal is being used it would smoke.

When *soft coal* is being hand fired it is necessary to use certain tools besides a shovel. You need a rake, Fig. 5-25; a hoe, Fig. 5-26; and a slice bar, Fig. 5-27. The sizes given in the sketches are suggested by the Bureau of Mines. It is necessary to move and work the firebed when burning soft coal as the burning coal sticks together. This helps the air to get through the grates and aids in combustion.

When firing *hard coal* by hand the fuel bed must not be disturbed. You must only keep the fire level and the holes covered up. If you disturb the firebed using hand tools you will cause the fire to drop through the holes in the grate and you will lose the fire.

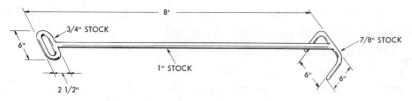

Fig. 5-25. Rake for soft coal firing by hand.

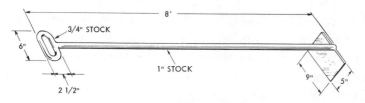

Fig. 5-26. Hoe for cleaning coal fires.

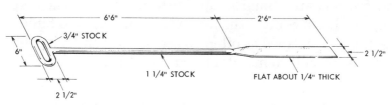

Fig. 5-27. Slice bar for breaking up coal fire bed.

CLEANING A FIRE

Hand fired boilers may have stationary grates (these are grates that do not move) or they may have dumping grates. Grates that can be opened dumping the ashes on the grates into the ash pit.

When cleaning fires on stationary grates the fireman must pull the ashes and clinkers out either onto the floor or into a wheelbarrow. The fires must be cleaned completely and quickly. If they aren't cleaned completely the result will be a dirty fire with ashes and clinkers still present. If they aren't cleaned quickly the steam pressure will drop.

There are two methods used in cleaning fires. One is to pull much of the clean fire to one side. Burn down the side to be cleaned so you will lose as little good coal as possible. Then use a hoe to pull out all the ash and clinkers. When the grate is completely clean pull over the clean fire from the other side and coal over lightly. Build the cleaned side up, burn the other side down following the same procedure as before.

The second method is to pull the actively burning coal to the front of the firebox and remove the ash and clinkers from the back by pulling it over the top of the burning coal. The clean fire is pushed back and the front cleaned. Coal over lightly and build up the fire.

In order to clean a boiler with dumping grates you push the good fire to the back, dump the front grates and then pull the good fire forward. Build it up, burn down the back and then dump the back. Spread the fire evenly and start to build it up. Fires must be cleaned as often as necessary. When they become dirty it is difficult to maintain your plant load.

Looking Back

1. Hand fired coal should be spread evenly and lightly over the fire bed. This prevents smoke and slowing down of the fire.

2. Keep fires clean by removing ashes and clinkers as often as necessary. Dirty fires do not heat rapidly.

3. Clean fires in one half of the firebox and then pull the fire over to the clean side and clean the other.

4. Smoke is a sign of incomplete burning of the coal. Too much coal at one time, insufficient air, or improper firing are the likely causes.

5. Break up the fire bed when using soft coal. Use a slice bar or similar tool to open up the coked mass and let air through.

STOKERS

Stokers were introduced to increase the efficiency of burning coal. They also made it possible to burn more coal in the same size boiler and made it possible to build larger coal burning boilers. The purpose of a stoker is to introduce the coal into the firebox or furnace of the boiler evenly. It eliminates the need of opening the fire door to coal over. This keeps the furnace at the same hot temperature and cuts down on chilling the brick work. The common type of stoker used in low pressure plants is the underfeed type. Boilers will have one or more depending on the size of the boiler and the plant load.

Stokers consist of a hopper to hold the coal and a mechanism to feed the coal into the boiler furnace. A fan to provide air for burning the coal is a part of the stoker. There are two types of underfeed stokers which differ in their method of feeding coal. One is the screw feed while the other is the ram feed type.

Screw feed stoker. The screw feed stoker, Fig. 5–28, has a hopper that is designed for easy filling by hand and to assure a free flow of coal. Side extensions and a large front apron provide extra large capacity. There is a cut-off gate at the bottom which permits removal of foreign material, such as stones, from the coal feed mechanism without emptying the hopper.

A short, low speed feed worm conveys the coal from the hopper to the retort. The worm is designed to prevent the coal packing. It is located entirely outside the fire zone and is free floating in the conveyor housing. The housing has an access door for inspection or removal of obstructions.

The retort has an expanding mouth which prevents packing of the coal and insures an easy flow of coal on to the grates. The retort is the full length of the stoker and provides maximum underfeed combustion area. The retort is located in the main air chamber and is kept cool by the entering air. The retort is anchored only at the front so it is free to expand and contract. A coal feed pusher is located in the bottom of the retort and is reciprocating (moving forward and back). It can be adjusted to

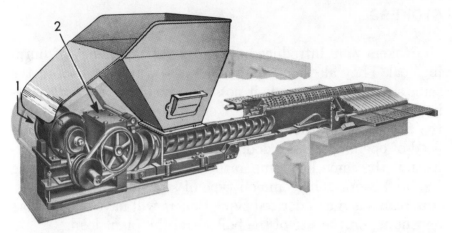

Fig. 5-28. Coal stoker showing screw feed mechanism. Combustion Engineering Co.

distribute the coal evenly from front to rear and to provide proper agitation to maintain a porous, level fuel bed through which the air may pass with little resistance to maintain combustion.

The grate bars have an alternating arrangement of moving and stationary bars. The moving bars have a short side motion which in combination with the reciprocating coal pushers in the retort provide agitation and even distribution of fuel over the entire grate area. This also causes a gradual movement of the burned out refuse (ash) to the side dumping grates.

Ash dumping grates are located at the furnace side walls at a point farthest away from the entering coal and the hottest combustion zone. The dumping grates are made in sections and are mounted on a heavy bar that is connected to a hand lever at the front of the stoker. These grates are provided with a retaining ledge to hold the fuel bed while the grates are being dropped to dump the ashes.

The stoker drive motor operates through a variable speed transmission which provides for changes in the rate of coal feed. The transmission is enclosed and runs in oil. There is a shear key or pin somewhere in the transmission that prevents damage to the stoker in case of an obstruction clogging the feed screw. The shear key is easily accessible and easily replaced. The stoker drive mechanism also has a clutch that may be used to stop the coal feed while the fan continues to operate. This permits complete burning down of the fire when banking the boiler or taking it off the line. It also aids in starting a fire by providing air while the coal is being ignited.

The forced draft fan supplies air for combustion. It is directly connected to the drive motor. The fan housing is connected to the stoker housing and the volume of air is regulated by a damper on the fan inlet.

There are three combustion zones that provide for complete combustion over the entire grate area.

1. An underfeed zone at the retort where the volatile gases are driven off as the coal is heated and where they are burned in the furnace.

2. Moving grate zone where the coal is burned.

3. Dumping grate zone where any remaining coal may be burned before discharging to the ash pit.

Overfire air is provided under separate damper control for complete combustion and smoke prevention.

Ram feed stoker. The ram feed stoker, Fig. 5–29, uses a feeder block ram instead of a screw to feed the coal into the retort. The ram can be driven by an electric motor or by a steam piston.

This stoker, like the screw feed stoker, appeals to operating firemen because of the small amount of attention required to maintain a good fire and efficient combustion over a wide range. This is due to the uniform feed and even distribution of the coal, quick discharge of ash without disturbing the firebed, and proper air distribution and control.

The coal feed ram moves the coal in the hopper into the retort by a reciprocating, sliding bottom on which are mounted a feeder block or ram at the hopper end and auxiliary pushers at intervals in the retort, Fig. 5–30. With each inward stroke the feeder block forces a definite quantity of coal into the retort.

Fig. 5-29. Coal stoker showing ram feed. Combustion Engineering Co.

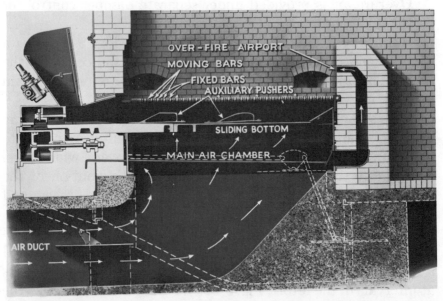

Fig. 5-30. Moving grate bars underfeed stoker. Combustion Engineering Co.

The auxiliary pushers aid in the distribution of the fuel in the retort and agitate the fuel over the retort.

As the inward stroke progresses the flow of coal from the hopper is cut off by the feeder block. When the outward stroke occurs coal falls in front of the block to be pushed into the retort on the next stroke. The amount of coal fed depends on the frequency of the strokes. The length of the stroke remains the same regardless of the rate of coal feed.

Coal distribution is similar to that in the screw feed stoker. The grate bars are alternately moving and stationary causing a gradual movement of the coal toward the dumping grates. By properly adjusting the frequency of the strokes to the load conditions the fuel is burned before any part of it reaches the ash discharge (dumping) grates.

A sliding retort carries the main ram and auxiliary pushers which feeds coal from the hopper to the rear and at the same time agitates the fuel in the retort.

The drive may be an integral steam piston, hydraulic piston, or gear transmission. The rate of fuel feed is controlled by

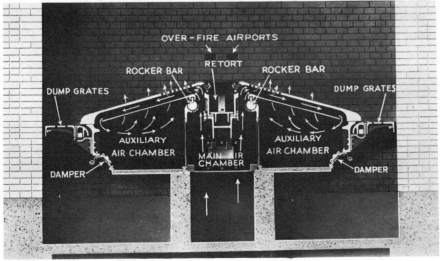

Fig. 5-31. Air distribution and dumping grates of stoker. Combustion Engineering Co.

timing the intervals of the strokes not by varying the length of the stroke as a constant length of stroke reduces the tendency of the fuel bed to cake at low operations and provides adequate agitation of the fuel bed.

Coal distribution is assured by a large throat opening to the retort, full stroke of the ram and pushers, and the sliding bottom. The sideways motion of the grates causes the coal to move gradually toward the dumping grates.

Air distribution, Fig. 5–31, is designed so that the air pressure is greatest where the fuel is thickest, over the retort.

The air control has three forced draft zones parallel to the retort. There is separate control of the air to the grate surfaces, the dumping grates, and over the fire.

STARTING A STOKER

When starting up a stoker you first inspect the hopper to make sure it is free of any foreign material. Any large piece of stone or metal could cause the shear pin or key to break. A shear pin is provided so that you will not damage the driving mechanism, the screw, or ram feed. Then the stoker should be greased and the oil level in the gear box checked. Now you can fill the hopper and start running the coal into the retort. The coal should be fed in until the grates are covered. Then wood is laid on top of the coal and waste rags added. Under no circumstance should a volatile liquid be poured on top of the wood. The stoker should be disengaged once enough coal has been put into the boiler. The outlet damper should be cracked open and the rags lighted. The forced draft fan should be running with only a small amount of air admitted. Once the coal starts to burn the stoker feed can be started very slowly, air increased and the boiler should be warmed up slowly.

Banking. When the fire is to be banked the coal feed is disengaged and the fire bed burned down very low. Then the forced draft is shut off and the coal run into the retort until only a small ribbon of fire exists along the whole length of the grates. The outlet damper is cracked open slightly and the forced draft left off. Periodically the stoker must be run to introduce green coal into the retort. You must never allow the fire to burn down into the retort. This would damage the auxiliary pusher blocks. Some stokers are equipped with an automatic banking control. This control will start the stoker every half hour and allow it to run from one to ten minutes depending on how it is set. This keeps the live fire from damaging the retort, and keeps the fuel bed burning.

Burning down a fire. When the boiler is to be taken off the line you must start early so that you can burn all the coal that is in the hopper. After the hopper is empty it is filled with ashes. The ashes will fill the retort pushing out all the coal onto the grates where it can be burned. The forced draft should be cut down to allow the firebox to cool slowly. The water level should be watched and the boiler valves handled in the same manner as previously described. After the boiler has cooled enough all the ashes and coal must be cleaned from the retort and grates.

Looking Back

1. Stokers feed coal and air into a boiler where they are burned. The coal feed rate and the air volume are controlled for best combustion.
2. There are controls for regulating the coal and air. Adjust these according to the fire wanted.
3. To start a fire fill the retort with coal and then build a fire of wood and rags on top.
4. Never use kerosene, gasoline, or other fluids for starting a fire.
5. In shutting down a stoker be sure to burn all the coal in it and then clean the hopper and retort thoroughly.

COMBUSTION

It is necessary for you to become familiar with a number of terms commonly used in connection with combustion of fuel.

1. *Combustion* is the rapid burning of fuel and oxygen. It takes between 14 and 15 pounds of air to burn a pound of oil.

2. *Perfect combustion* is the burning of all the fuel with exactly the correct amount of air. This can only be accomplished in a laboratory under carefully controlled conditions.

3. *Complete combustion* is the burning of all the fuel with the proper amount of excess air. This is what we strive for in the field.

4. *Incomplete combustion* is a condition which exists when all the fuel is not burned. The result is smoke and soot.

There are four things needed for complete combustion. The key word "MATT" will help you to remember them.

M — Mixture of fuel and air must be right or the air-fuel ratio must be proper.

A — Atomization or the breaking up of oil into particles that are surrounded by air.

T — Temperature of the fuel must be right to support combustion or at the fire point of the fuel.

T — Time to complete combustion within the firing chamber.

Note that if combustion is not completed before gases come in contact with the heating surface (the part of the boiler that has water on one side and gases of combustion on the other side) the gases will cool and soot and smoke will result.

There are two types of air we refer to when we speak about combustion:

1. Primary air which is controlled by the number of gallons of oil you burn.

2. Secondary air which controls the efficiency of combustion.

It is the fireman's responsibility to control combustion so that there is no smoke, thereby saving on fuel cost and preventing air pollution. If soot builds up on the heating surface it acts as insulation and prevents the transfer of heat to the water. This causes the temperature of the gases being discharged to the smoke stack to rise. In order to help the fireman, the boilers are usually equipped with combustion controls.

COMBUSTION CONTROLS

The purpose of combustion controls is to help the fireman maintain a proper air fuel mixture and control the burning rate of the fuel. High fire is burning the maximum amount of fuel; low fire is burning the minimum amount of fuel.

The combustion control used on low pressure heating boilers is the "on" and "off" type. This means that when the steam pressure drops to a certain pre-set pressure the boiler will start and when the steam pressure rises to its pre-set pressure the boiler shuts off. Thus the name "on and off" type of combustion control describes it very well.

There are three things that a combustion control must regulate:

1. Fuel supply in proportion to steam demand.
2. Air supply.
3. Ratio of air to the fuel supplied.

Let's see what is meant by each one.

1. Fuel supply in proportion to steam demand. If it's a cold day more steam will be needed to heat the building. (The boiler must carry a heavier load.) In order to do this it is necessary to burn more fuel or carry a high fire. As the load or demand for steam drops off the boiler carries a lighter load or goes into low fire.

2. Air supply. Both primary and secondary air are needed to burn fuel. Fig. 5–11 shows that a fan is mounted on the same shaft as the rotary cup of the burner. This supplies primary air. The primary air is controlled by the amount of oil that is being burned. The secondary air controls the efficiency of burning the oil and is usually introduced into the firebox from below the burner.

3. Ratio of air to the fuel supplied. The more fuel you burn the more primary and secondary air you need. As the steam demand increases the boiler goes into high fire, the oil valve is opened through linkages and allows more oil to flow to the burner. At the same time a damper is opened and more primary

air flows to the burner to mix with the oil. The linkage also opens the secondary air damper introducing more air to complete the combustion. As the load drops off the oil valve closes throttling the flow of oil to the burner. At the same time it cuts down on the primary and secondary air.

Watch your boiler the next time it lights off. Watch the linkage move. See how it opens the secondary air damper and listen to the fire as it goes into a high fire condition. You can see it happen and you can hear it happen also. A fireman who knows his plant can tell by the sound of the burner whether it is on high fire or low.

Looking Back

1. A burner should always light off in low fire and shut off in low fire. This is the job of the modulating pressuretrol discussed in Chapter 2.
2. The starting and stopping of the boiler on a pressure drop is controlled by the pressuretrol.
3. The pressuretrol and the modulating pressuretrol are both located at the highest part of the steam side of the boiler.

AUTOMATIC CONTROLS

Boilers burning #4, #5, or #6 oil use a gas pilot to light the oil. The pilot is ignited by a spark plug which is energized by an ignition transformer. Boilers burning #2 oil use electrodes that are energized by an ignition transformer to light the oil. Since #2 oil, which is the type used in homes, is expensive and rarely used in larger boilers, we will confine our discussion to the heavy oils that are most frequently used.

Fig. 5–34 shows the way the ignition spark lights the gas pilot which lights the oil. When the oil valve is energized it opens allowing oil to flow through the fuel tube assembly into the burner, and then into the combustion chamber.

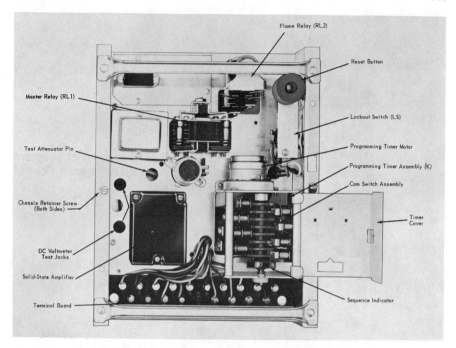

Fig. 5-32. Programming control clock.

The programming clock. This is the master mind, Fig. 5–32, that controls and directs the firing cycle. Before we discuss this control it is necessary that we understand a new term, purging.

Purging a firebox means to eliminate any unburned fuel in a gaseous condition that is in the firebox and that might ignite and cause a furnace explosion.

The firebox is hot. Any unburned fuel in the firebox would vaporize due to the heat.

This vapor or gas could ignite if exposed to an open flame. If a quantity of this gas should accumulate it would cause an explosion in the furnace that is just as destructive as a boiler explosion. To *purge* a firebox of these dangerous fumes, air is blown in and the gases are blown up the chimney or stack.

The programming clock has a shaft with fiber cams, Fig. 5–33. As the shaft rotates phosphor bronze leads containing

Fig. 5-33. Fiber cams of programming control. Electronic Control.

silver contacts ride on the cams and make electrical contacts at
the proper time. These contacts energize circuits that control
the operation of the burner in a definite program or sequence.
This control is similar to the one on an automatic washing
machine.

Firing sequence. Let's trace the firing cycle of a boiler with
the following settings: the operating range is 3 to 8 psi, the
modulating pressuretrol is set at 5 psi. When the steam pressure drops to 3 psi the pressuretrol completes an electric circuit to the programming clock which starts to turn. The first
contact on the first cam completes a circuit which starts up the
burner motor that rotates the fan and rotary cup. (See Fig.
5–11).

The fan blows into the fire box and purges it. This purging
takes about 30 seconds but may take as long as 60 according to
the type of burner and the programming control.

The programming clock is still turning and the second contact closes completing a circuit to the ignition transformer caus-

ing a spark in front of the gas tube. At the same time it opens a solenoid valve in the gas line allowing gas to flow through the tube and be ignited by the spark. The gas flame burns in front of the rotary cup.

The clock is still rotating and the next contact completes a circuit that opens the fuel oil solenoid valve allowing fuel oil to be pumped into the firebox. The oil is ignited by the gas flame.

The clock is still rotating and the cam that allowed the circuit to the gas and spark is opened. The circuit is broken and the solenoid gas valve is de-energized causing the gas valve to close. The spark stops and the burner is self-sustaining.

The program clock completes its cycle and then stops. The boiler is now under the control of the pressuretrol and modulating pressuretrol and the burner starts to go into high fire. The steam pressure starts to rise. When the pressure reaches 5 psi the modulating pressuretrol will cause the burner to go into a lower fire. If the pressure in the boiler continues to rise the burner will go into its lowest fire position until the pressure reaches the shut off point, which in our case is 8 psi. The pressuretrol then will shut off the burner in low fire. The program clock is started up again and the fan blows out gases left in the firebox. This is called a post purge. The clock stops after the post purge.

It is important for the fireman to understand the firing cycle of his boiler. He should time the cycle and know how long it takes the gas pilot to light and the oil valve to open. By doing this he can be sure that the boiler is operating properly.

In discussing the programming clock and its firing cycle one important feature, the fire eye, or flame detector, Fig. 5–34, was omitted in order to keep the operation simple.

The fire eye. This is a detector type of control that functions as a safety device in preventing furnace explosions. There are several types of flame detectors but the one used most often consists of a small lead sulfide cell, Fig. 5–35, that is placed in position so it can "see" the pilot light and main burner flame. The correct lining up of a fire eye is shown in Fig. 5–36.

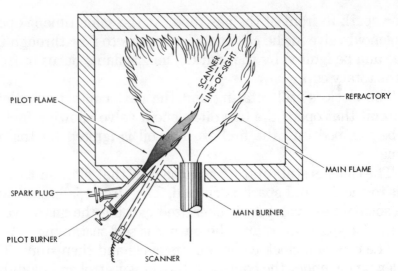

Fig. 5-34. Sketch of fire detector system. Pilot, main flame, and fire eye scanner sighting of both.

Before its operation is explained here is the story of how the lead sulfide cell was first used.

During World War II it was necessary to fly night missions to bomb strategic targets. The darkness gave American planes some protection from enemy fighter planes. However, the enemy soon was able to overcome the disadvantage of darkness and the loss of American night bombers increased sharply. No one knew how the enemy fighter planes could be so accurate in the dark until one was captured in good condition. Careful investigation disclosed small lead sulfide cells mounted in the wing tips. When the fighter pilot flew behind a bomber the lead sulfide cell picked up the light frequency of the engine exhausts and set off a signal on the control panel of the fighter plane. The pilot then knew he was in proper position and had only to pull the trigger.

It wasn't long after the war that the lead sulfide cell was used for the more rewarding job of saving lives by preventing furnace explosions. Let's see how it works.

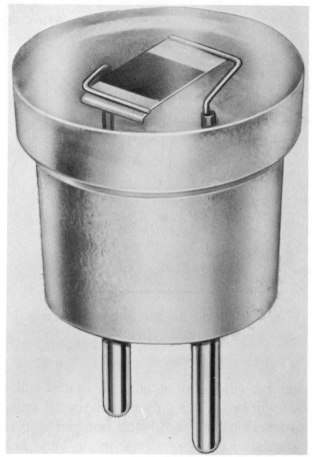

Fig. 5-35. Close up view of a lead sulphite cell. Electronic Control.

When the pressuretrol calls for steam the program clock starts the purge cycle and then the gas pilot lights. The fire eye "sees" the pilot light (this is called proving the pilot) and allows the cycle to continue. The oil valve opens, the oil ignites, the fire eye "sees" the main flame (this is called proving the main flame) and holds the circuit in and the program clock continues carrying out its prescribed functions.

If the fire eye scanner does not prove the pilot during the

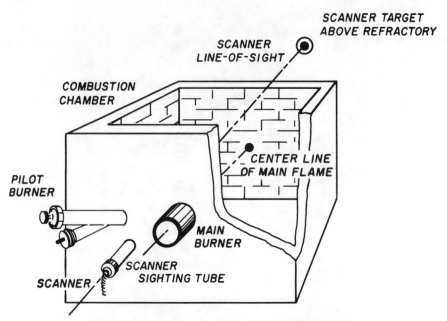

Fig. 5-36. Proper position of fire eye scanner.

pilot proving period the pilot solenoid valve will be de-energized. This will break the circuit and the oil valve will not open, thus preventing oil from being pumped into the firebox without a source of ignition. If the main flame fails the fire eye will also shut off the oil valve, putting the burner into a flame failure condition. This also prevents a build up of oil in the firebox. The only way the cycle can be started again is by pushing a reset button in the programming clock control.

The fireman should be familiar with the part the flame safeguard and programming control systems play in the firing cycle.

The company that supplies the program clock has complete bulletins on its equipment. These include wiring diagrams and charts indicating where to look for failures and how to correct them. They also have suggested replacement and testing schedules. This is all part of the service offered when you use their controls. Take advantage of it. These companies are reliable and all try to produce the best control on the market. However, good service is important. The company whose field engineers answer your questions and supply you with data sheets and maintenance bulletins are the ones to remember. The best companies not only put out good products but they back them up with good service.

Looking Back

1. Be sure a boiler is purged of all fumes or gas after each firing cycle.
2. The fire eye detects any trouble in the lighting of the pilot light or the burner.
3. There is a definite time sequence in the operation of the flame safeguard and the programming systems.

6

DRAFT CONTROL

We learned in the first chapter that a fire needs air to burn. In burning, the oxygen in the air combines with carbon in the fuel and generates heat in the process. The heat is in the gases of combustion and they must contact the heating surfaces to give up or transfer their heat to water and cause it to boil into steam.

To have combustion air has to be brought into contact with the fuel at the right temperature. Then the hot gases have to be moved away from the fire to the place where the heat can be transferred, the heating surfaces or boiler tubes, and the gases have to be removed to make room for more. In short, there has to be a flow of air into the fuel and then a flow of hot gases to the heating surfaces and then they must be moved out of the way.

To have a flow there has to be a difference in pressure between two places. The flow goes from the place that has the higher pressure to the place that has the lower pressure. This movement or flow in air or gases is called a draft.

There are two ways of creating a draft in a boiler. One is by arranging the boiler so that a flow is caused by the natural action of the air or the gases. The other way is to use some type of machinery for making a draft. These are termed natural and mechanical drafts.

Natural drafts are not used with automatic fuel systems as they cannot be as precisely or rapidly regulated.

NATURAL DRAFT

A natural draft uses one of the basic laws of physics. That is that warm air rises. Air as it is warmed expands and becomes lighter in weight. Colder, heavier air pushes in under it and forces it up. This causes a draft. By directing this draft into a pipe we can control it and increase or decrease its speed. Natural drafts are used in many places, especially with hand firing. Fireplaces and stoves use natural drafts to make them burn. Some small boilers also use natural draft.

Natural draft is regulated by controlling the air and heated gas. The more a gas is heated the faster it rises. Also a column of heated gas will weigh less than a column of cool air so the higher the column of gas the more speed it will develop as the weight of the column of cold air increases. If we have a chimney 10 feet high and burn some paper in the bottom of it to heat the air in the chimney the air will rise and cold air will rush in at the bottom to take its place. Fig. 6–1.

If we increase the height of the chimney we increase the column of warm air or gas and more air will rush in the bottom to take its place and this will create a stronger draft. If the chimney is closed at the bottom and the only way for the cold air to get in is through a pipe we can control the amount of air and the direction. We can make air come through a boiler and the breeching or smoke pipe to get to the chimney.

If the air has to come through burning fuel we make a hotter fire by providing more air for the fuel to use. If we let the air go into the chimney without going through the fuel we can cut down on the amount of air the fuel is getting and that will keep the fire smaller. So natural draft is regulated by the amount of the heated gases going up the chimney and the height of the chimney.

By adding cold air to the gases going up the chimney their heat is reduced and the movement is slower. Dampers or draft regulators placed after the fire let air go directly into the chimney and dilute the heated gases, slowing them down and reducing the amount of air being pulled into the fuel.

WARM AIR

10'

COLD AIR

FIRE

Fig. 6-1. Principles of natural draft. Comparison of columns of cold and warm air.

If the dampers are closed the air has to come through the fuel and the rate of combustion is increased causing the fire to become hotter and the gases to rise faster creating more draft. Fig. 6–2.

MECHANICAL DRAFTS

There are two types of mechanical drafts: forced and induced. These two types can be understood by comparing them

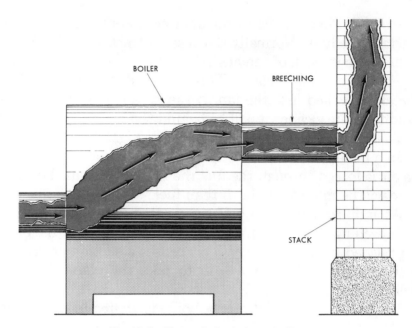

Fig. 6-2. Natural draft in a boiler.

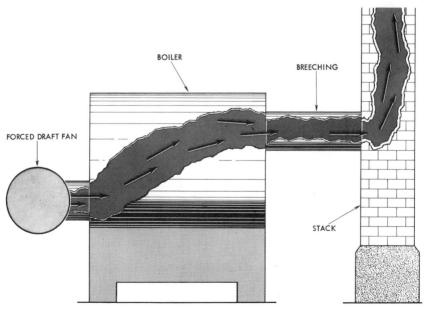

Fig. 6-3. Forced draft in a boiler.

to a vacuum cleaner. Most vacuum cleaners have two places to attach the hose. Normally the hose is attached to the cleaner so the fan is in back of the hose. This sucks the air through the hose and creates a vacuum. This is an induced draft-suction. The air is sucked into the fan. If the hose is put on the other opening in the cleaner it will be behind the fan and the air will be blown out of it. This is a forced draft. Fig. 6–3.

In a boiler the air can be forced into a firebox by a fan. If the air is forced through the fuel the fire will become hotter as there will be more air to burn than fuel.

An induced draft in a boiler is created by putting a fan in the breeching or smoke pipe so the air is pulled out of the boiler and pushed up the chimney. This causes a suction or vacuum in the firebox and air moves in to take the place of the air sucked out, Fig. 6–4.

When using natural draft the boiler is limited in the amount

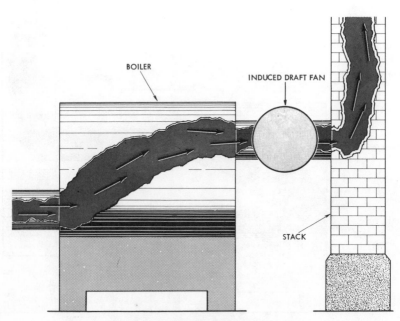

Fig. 6-4. Induced draft in a boiler.

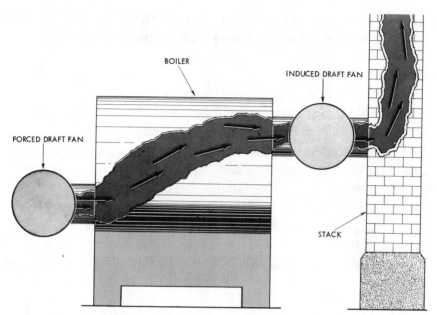

Fig. 6-5. Combination forced and induced draft system in a boiler.

of fuel that can be burned as there is no complete control over the draft. In winter a higher draft is possible because of the difference in the temperature inside and outside the chimney or stack. In summer this same stack cannot produce as high a draft. With mechanical drafts the amount of air can be controlled and more fuel can be burned to produce more steam.

It should be noted that with forced draft a good chimney is needed to carry off the hot gases from the firebox. With induced draft this is not necessary as the gases are being blown into the chimney. Consequently induced draft will overcome poor chimney design that forced draft cannot. In some systems both forced and induced drafts are used. Fig. 6–5.

We have talked about draft and its effect on boiler operation. We have also talked about steam gages and the purpose they serve. You remember that a steam gage registers pressure in pounds per square inch. Draft gages are calibrated differently. The pressure they measure is so small it is measured in inches or tenths of inches of a vertical column of water.

MEASURING DRAFT

A simple draft gage, called a manometer, is shown in Fig. 6–6. It consists of a bent glass tube shaped like a U. One end of the U-tube is open and the other end is connected to a flexible hose. The U-tube is partly filled with water until both legs have the water level at 0. The flexible hose is placed at the point where we wish to measure the draft such as the boiler breeching. If the pressure in the breeching is less than in the atmosphere it will suck some of the air out of the hose and the water will be sucked up in the side of the tube connected to the hose

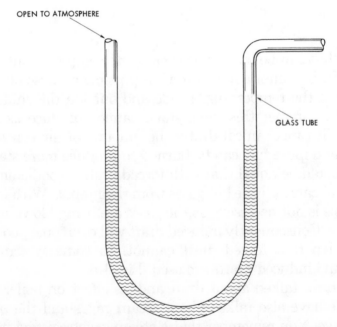

Fig. 6-6. Manometer measures draft with U-tube type gage. Both legs are equal level at atmospheric pressure.

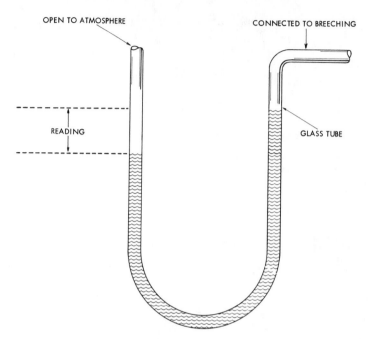

Fig. 6-7. Draft gage connected to breeching showing negative reading. Water rises in leg connected to the breeching.

and that will cause it to drop in the other side. Fig. 6–7. If the pressure in the breeching were higher than outside there would be a push against the water in the tube and it would rise in the open side of the U-tube. Fig. 6–8.

If the pressure in the breeching is less than the atmosphere it is termed a negative reading. If the pressure is higher it is termed a positive reading.

CONTROLLING DRAFT

The fireman will have to learn the ways in which to control the draft in the equipment he is firing.

With hand firing the fireman will have to adjust the open-

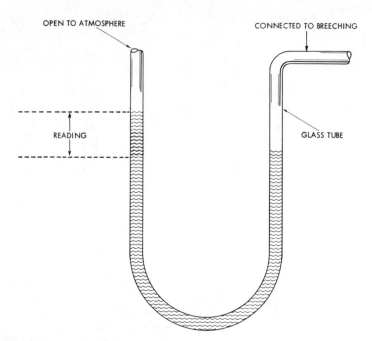

OPEN TO ATMOSPHERE

CONNECTED TO BREECHING

READING

GLASS TUBE

Fig. 6-8. Draft gage showing positive pressure in breeching. Water rises in opposite leg of tube than in Fig. 6-7.

ings in the ashpit door and the firedoor and the dampers in the smokepipe or breeching. This will take a little practice working the equipment and learning the details from someone familiar with the plant is desirable.

Where oil, gas, or stokers are used there are certain to be controls for regulating the draft. The manufacturers of the equipment provide instructions for operating these controls and the field representatives of these companies should be contacted for any questions concerning the adjustment or operation of them.

The fireman should remember that the control of draft will do much to control the quality of combustion so that smoke and air pollution are kept to a minimum and the maximum amount of heat is produced from the fuel. Too much draft may cause too hot a fire with coal or may remove the gases from the firebox before they have had an opportunity to transfer their heat to the water. In short, the heat may blow up the stack. Too little draft will not provide enough air to burn all the fuel and some of it may go up the stack as carbon particles or soot and smoke. In either case the result of incorrect draft will be wasted fuel and possibly other problems.

There are several variations of draft and the fireman should learn to recognize each kind.

Looking Back

1. Draft is a movement of air or gas due to uneven pressure. In a boiler the draft moves from the outside through the firebox to the stack or chimney.
2. The amount of draft determines the rate of combustion or burning. Increasing the draft will burn more fuel and create more steam.
3. There are two kinds of draft. One is natural and is caused by warming the air inside the boiler so that it rises and goes up the stack. The other is mechanical and is made by fans.
4. Draft must be regulated to provide enough air for the proper burning of the fuel. Too much draft wastes heat up the stack; too little doesn't burn all the fuel and makes smoke.
5. Mechanical drafts may be either induced or forced or a combination of both.

7

WATER TREATMENT

All city water and most water from wells contains minerals. These minerals or salts as they are called chemically do not cause much trouble as long as they are dissolved in water—in solution. However, water can only hold so much of these salts and when they become concentrated the salts settle out of the water and form mineral deposits or scale.

Water that contains large amounts of minerals is called hard water. Water that has small quantities is called soft water.

When water is turned to steam the minerals are left behind so the water in the boiler becomes saturated with the minerals and they settle out of the water and form a scale on the boiler surfaces. This is the same thing that happens in a teakettle or any container used for boiling or evaporating water. Scale causes two problems in a boiler. First, it acts like insulation and slows down the transfer of heat to the water. This means you have to burn more fuel to make steam. Second, if the water cannot take away the heat, the heating surface overheats and blisters and bags will form in the metal and burned out tubes will result.

If you were to take a boiler tube and place it in the firebox of a boiler that is being fired the tube would bend like a pretzel in a very short time. Yet the same tube in the boiler with water passing around or through it (depending on the type of boiler) is not injured. Why? Because the water is removing the heat and keeping the tube from overheating. As long as there is nothing but the tube between the water and the gases of combustion this transfer of heat takes place, and there is no trouble. However, if scale builds up on the heating surface the scale insulates the tube from the water and the tube overheats and burns out.

Scale also narrows the inside of the tube and slows down the circulation of water through it. This, with the insulating effect of scale, lets the metal of the tube overheat.

PREVENTING SCALE FORMATION

This is where feed water treatment enters the picture. We cannot destroy the minerals in the water but we can change them so they do not settle out and form scale. By adding chemicals to the boiler water the scale-forming salts or minerals are changed to ones that make a non-adhering sludge; a mud that stays in the water and that does not settle out. Now we don't have scale to worry about but we do have mud to get rid of. How is this done? In a previous chapter you learned about Bottom Blow Down Lines. They are located at the lowest part of the water side of a boiler. All we have to do is get the mud to the bottom of the boiler and we can blow it out.

If you have ever watched water boiling in a kettle you know it is moving rapidly when it is boiling and as it is removed from the heat it calms down. When a boiler is under a heavy load it is boiling vigorously and the water is circulating rapidly. As the the load drops off the circulation slows down and the mud settles to the lower part of the boiler. Just as the vegetables in a pot of soup will be rapidly circulating all through the kettle when it is boiling hard so will the mud in the boiler. When the boiling stops or slows down the vegetables settle to the bottom of the kettle and, similarly, the mud settles to the bottom of the boiler. If we open a valve at the bottom of the boiler then the mud will be blown out. This is what we do with the Bottom Blow Down Line. This is opened regularly during a time when the boiler is on light load and the mud is blown out.

PRIMING AND CARRY-OVER

These two problems are similar and have the same causes and cures. Priming is the problem of small particles of water being carried into the steam lines. Carry-over is the problem where large slugs of water get into the lines.

Both of these can be caused by too high a water level in the boiler, by a high concentration of chemicals in the water, impurities in the water that cause a high surface tension, and by opening a main steam valve too fast.

Both can cause water hammer and possibly a bursting of the steam lines.

Using a bottom blow will reduce the water level and also decrease the concentration of chemicals. A surface blow may be needed if the water is foaming. This valve is located on the boiler directly on the normal water level. When it is opened a mixture of steam and hot water will shoot out. This will carry off any impurities that are floating on the surface causing the high surface tension. Care in opening the steam valves will prevent carry-over from sudden pressure changes.

A bottom blow may be needed to reduce the water level and to correct for high concentration of chemicals. Where there are too many chemicals the bottom blow should be followed by the addition of make-up water to dilute the chemicals to a safe level.

High surface tension is shown by a rapid fluctuation of the level in the gage glass. The glass will fill and empty rapidly indicating a foaming condition in the boiler.

OXYGEN DANGER

The fireman has to be careful of scale-forming salts in the water and he has to be careful of oxygen in the water too. The free oxygen in the water is liberated when the water is heated. The air containing the oxygen is driven from the water. The oxygen causes corrosion (rusting) and pitting of the boiler metal. This problem is handled in two ways. One is to heat the water before it is put in the boiler. This will drive off some of the air. The other way is to add chemicals that combine with the oxygen to make it into a harmless compound. These chemicals are called oxygen scavengers. One of them that is used a great deal is sodium sulfite.

FEEDING CHEMICALS

In low pressure operation chemicals for preventing scale and reducing oxygen are usually introduced into the boiler water by using a by-pass feeder, Fig. 7-1. This is installed on the discharge side of the boiler feed pump or vacuum pump.

The operation of the by-pass feeder is illustrated in Fig. 7-2. Valve number 1 is open and valves 2 and 4 are closed. Valve 3 is opened slowly to make sure there is no pressure in the feeder. There should be no pressure unless valves 2 or 4 are leaking. When valve 3 is open the chemicals are poured into the funnel. Then valve 3 is closed and valves 2 and 4 are opened and valve number 1 is slowly closed. The water has to flow through the chemical feeder forcing the chemicals into the boiler. This is sometimes referred to as slug feeding a boiler as the chemicals are moved into the boiler in a slug.

In some plants a more elaborate system of chemically feeding is used. This system has a tank in which the chemicals are mixed with water. A small pump on the bottom of the tank takes the chemical mixture from the tank and discharges it through pipe lines directly to the boiler. The pump is designed so its stroke may be adjusted. This permits controlling the amount of chemicals being fed to the boiler. This method can be used to feed the boiler continuously, taking as long as 24 hours to empty the tank.

WATER TREATMENT

Water treatment for most boilers is not too difficult. If the boiler gets back all or almost all of the condensate returns there is no need to treat the water. The only make-up water needed

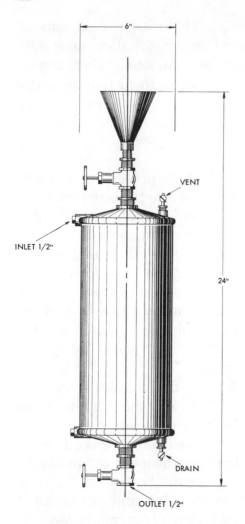

Fig. 7-1. A by-pass water
treatment feeder.

is that necessary to replace losses from blowing down the gage glass, water column, low water cut-off, and to take care of leaks. The less make-up water used the less scale forming salts will be introduced and the less chemicals will be needed. Less blow downs will be needed to remove the sludge.

Be sure to find out what chemicals have been used to treat the feed water and how often it has been done. The chemicals used to treat boiler water will depend on the minerals in the

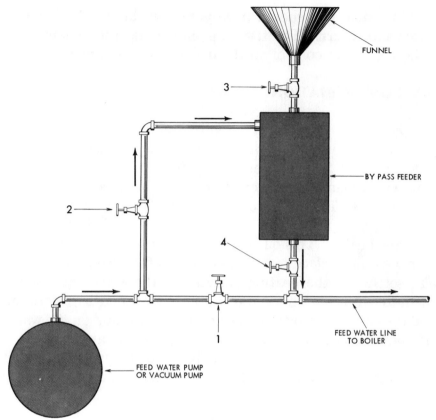

Fig. 7-2. Installation of by-pass feeder.

raw feed water and must be determined by tests. Find out what the practice has been with the boiler you are operating. If you do not know of a firm specializing in boiler water treatment contact the field man of the boiler manufacturer or the local boiler inspector.

There are many companies in the feed water treatment field. It would be wise to contact two or three and have them send their representatives. They will take samples of the water used for make up boiler water, and condensate returns. They will have the samples analyzed and then submit a proposal to cover your individual problems.

If possible you should have your own kit to check boiler water and returns. The tests are simple and they enable you to keep a better control on the boiler water treatment.

OIL CONTAMINATION

The danger of contamination of boiler water by oil occurs in plants that use No. 6 oil. It is necessary to heat this oil in order to pump it and to heat it again in order for it to burn. This was explained in the chapter on combustion.

It is possible for the steam heating coils to leak and for oil to get into the feed water system with the condensate returns from the heaters. This oil is very dangerous inside the boiler. It increases the surface tension and causes the water to foam. It also settles on the heating surfaces of the boiler and causes them to overheat. This can result in blisters and bags forming and the tubes burning out. If signs of oil are found on the water side of the boiler, it should be taken off the line at once.

It is necessary to boil out the boiler with caustic soda and then thoroughly wash it. If this happens consult a company dealing with feed water treatment. Their company engineers can give you the necessary help in cleaning the boiler.

Because of this danger it is a good practice to dump all fuel oil heater returns to waste. This prevents the boiler from becoming contaminated with oil.

Looking Back

1. All city water contains some minerals and scale-forming salts.
2. Water containing large amounts of minerals and scale-forming salts is referred to as hard water.
3. Water containing small amounts of minerals and scale-forming salts is referred to as soft water.
4. Chemicals added to the boiler turn minerals into a non-adhering sludge.

5. The non-adhering sludge is removed from the boiler using the bottom blow down valves.

6. Oxygen is removed from the boiler water by heating it before it enters the boiler and by adding an oxygen scavenger to the boiler.

7. The oxygen must be removed because it causes corrosion and pitting of boiler surfaces.

8. Priming is the carrying of small particles of water into the steam lines.

9. Carryover is large slugs of water being carried over into the steam lines.

10. Priming and carryover must be watched because they can cause water hammer and possible rupture of steam lines.

11. Oil in the steam and water side of a boiler can lead to overheating and burning out of heating surface.

12. Oil can be removed from the steam and water side of a boiler by boiling it out using caustic soda.

13. Feed water treatment is an important part of safe and efficient operation of a plant.

14. A reliable company should be brought in to analyze and prescribe the treatment needed for your particular plant.

8

BOILER OPERATIONS

The previous chapters gave you an understanding of some of the equipment you may find in the boiler room. There are no two boiler rooms alike. They all will vary in some respect. However, if you understand the basic equipment you can relate it to your own boiler room. You should check your plant carefully. Make a list of the types of pumps, boilers, and other equipment. Write down make, model numbers, capacities, and all the other information given on the plates fastened to the equipment. Write for maintenance manuals and instruction sheets for this equipment. Be sure to give the manufacturer the size, model number, and other information. It is impossible and impractical to give all the details in a book. These you must learn from the manufacturer. The basic operating procedures are the same, however, and they are the ones discussed in this chapter.

TAKING OVER A SHIFT

What is the first thing a fireman should do when he enters a boiler room to start his shift? The very first thing is to check the water level on all boilers that are on the line—generating steam. This is done by blowing down the gage glass. The water should enter the glass quickly when the gage glass blow down valve is closed. This would indicate clear lines, free of sludge, sediment, or scale build-up. Then he should blow down the water column and finally the low water cut-off. Be sure the boiler is firing when the low water cut-off is blown down. This will enable the fireman to check its operation. When the low water cut-off is blown down the boiler should shut off. When the low water blow down valve is closed the boiler should again light off.

The fireman then should check the steam pressure and the condition of the fires. Next all running auxiliaries (fuel oil pumps, water pumps, fan, burner) should be checked for proper temperature, pressure, and lubrication.

Let's review these points again:

1. Check water level of all boilers on the line by blowing down the gage glass, water column, and low water cut-off.

2. Check the steam pressure and the condition of the fires.

3. Check the running auxiliaries for proper temperature, pressure, and lubrication.

After these important preliminary checks have been made the fireman should have a set routine of jobs to perform. This routine will prevent forgetting to do any of them. A suggested routine would go:

1. Change over the fuel oil strainers and clean the one that has been in operation.

2. Shut down the burners one at a time and clean the burner tip or rotary cup depending on the type of burner in the plant.

3. Remove the fire eye scanner with the boiler firing. The boiler should shut off in a flame failure condition. Clean the fire eye, replace it and re-set the program clock. Then watch the boiler go through a firing cycle and light off.

After making these safety checks and performing the routine duties mentioned the fireman can feel fairly sure the boiler will operate safely.

He can feel quite certain there is nothing seriously wrong with the fuel system, the feed water system, or the combustion control system, that would affect safe boiler operation. However, he should also know when the safety valve was last checked and the accuracy of the gages.

STARTING PROCEDURE

When you are starting up a boiler make sure that everything is in proper working condition. Check the following list to be sure you have remembered everything.

1. Check water level in the boiler.
2. Check the main stop valve. Be sure it is closed.
3. Check the fire side. Be sure there are no tools or rags left inside and no trace of fuel oil, gas, or fumes.
4. Open the air cock or the top try cock to vent the air from the boiler during the warm up.
5. Start the boiler up and keep it in low fire, if possible, in order to warm the boiler slowly.
6. Watch the water level as the boiler is being warmed.
7. When the steam gage records a pressure on the boiler, blow down the gage glass, water column, and low water cut-off.
8. Test the fire eye to be certain of its operation.
9. When the boiler is few pounds below the header pressure slowly crack the main boiler stop valve and allow the pressure to equalize. Then open the main boiler stop valve slowly until it is wide open. The boiler is now cut in on the line.

Note. The boiler should always be a little below the header pressure when cutting in on the line. This will prevent carryover because steam will flow into the boiler for a short time.

SHUT DOWN PROCEDURE

When taking a boiler off the line make sure the remaining boilers can carry the load. Then proceed as follows:

1. Secure (shut off) the fire on the boiler that is coming off the line.
2. Cut down on any forced draft to prevent the brick work in the firebox from cooling too rapidly. This will prevent spalling (flaking of the surface) from developing in the brick work.
3. Watch the water level.
4. When the boiler has stopped steaming close the main steam stop valve.

5. When the steam gage shows about 1 or 2 lbs pressure open the air cock or top try cock to prevent a vacuum from forming.

Note: Do not close the main steam valve immediately. If you do the safety valve will pop. The boiler firebox is hot and the heat in brickwork and ducts will continue to keep the boiler steaming for a short time.

Looking Back

1. Check water levels, steam pressures, fires, and running auxiliaries immediately when you come on duty.
2. Clean strainers, burners, and other controls and equipment.
3. When starting a boiler follow the starting procedure by checking water level, steam valves, air vents, and controls.
4. Shut down a boiler by following the sequence of cutting off the fire, stopping the draft, and watching the steam until the valves can be closed to take it off the line.

BLOWING DOWN A BOILER

It was pointed out that there are four reasons for giving a boiler a bottom blow down.

1. Blowing out sludge and sediment.
2. Controlling high water.
3. Controlling high chemical concentrations.
4. Dumping (emptying) a boiler for cleaning, inspection, or repairs.

There will be pressure on the boiler at all times when it is blowing down except when being dumped. Also the boiler must be cooled before it is dumped as dumping a hot boiler would result in damage to the boiler.

There are certain precautions a fireman must follow when blowing down a boiler.

1. Always check the water level before starting to blow down.

2. Try to blow down when the boiler is on a light load.

3. Never walk away from an open blow down valve. Keep your hand on the valve until it is closed.

4. Open the blow down valve slowly. Open it fully and then close it slowly.

5. Never keep the blow down open long enough for the water level in the boiler to drop out of sight in the gage glass.

Some boilers are equipped with two valves on the blow down line. Often they are a quick closing valve and screw type valve. The ASME code states that the quick closing valve should be the one closest to the boiler. Fig. 2–8. The quick closing valve is a sealing valve. The screw type valve is the blowing valve. It takes all the wear and tear of blowing down. It is designed to take this abuse. The quick opening valve should be opened first and closed last.

HANDLING LOW WATER

Just what is meant by low water? How do you know if it is safe to add water to the boiler?

After you have been operating a boiler for a while you will know just where the normal operating water level is in your boiler. It varies in different boilers. Generally it is about a half gage glass. As the water level starts to fall below this point the boiler is starting to develop a low water condition. If you can see water in the gage glass or if you get water from the bottom try cock when you open it you can add water to the boiler safely. Remember the lowest visible part of the gage glass must be 2–3 inches above the highest heating surface. As long as the heating surface is covered with water it is safe to add water to the boiler. But if you cannot see water in the gage glass or if you get steam from the bottom try cock *do not attempt to add water to the boiler*. The boiler must be secured, cooled slowly, and the boiler inspector notified. He will then inspect the boiler to see if there has been any damage due to overheating. This is the same procedure you would follow in the event the boiler dropped its fusible plug.

Never add water to a boiler if you cannot see water in the gage glass or if you get steam out of the bottom try cock when you open it. Adding water could cause a boiler explosion!

FURNACE EXPLOSIONS

A furnace explosion is the result of a build-up of combustible gases in the firebox. This build-up could be the result of a leaking gas or oil valve, or a flame failure not handled properly.

Let's see how we can avoid the chance of a furnace explosion.

1. If any odor of gas is detected the boiler should be shut down and not allowed to light off. All gas lines should be checked. This is done by going over every fitting with a mixture of soapy water and a paint brush. Gas leaks will show up as bubbles. The leak must be repaired.

2. Before lighting off a boiler the firebox should be checked for signs of oil. If oil is found it must be cleaned up and the source of it found and repaired.

3. In the event of a flame failure where the fire eye has shut off the boiler the firebox must be examined for signs of oil before lighting it off. Then the firebox must be purged. Purging a firebox means to get rid of the gases that might be present. You must know how to purge your own boiler. One method is to shut off the gas to the pilot, disconnect the spark from the ignition transformer, and shut off the oil valve. Then start the program clock and let the clock go through a cycle. The burner motor will run and the fan will blow out any build-up of gases. Then open the oil valve, reconnect the spark, open the gas valve and allow the burner to go through a firing cycle.

Do not turn the program clock ahead in an attempt to by-pass a purge cycle in order to start a burner that has misfired. You could cause a further build-up of raw oil in the firebox and this oil, as it gasifies, could lead to a serious furnace explosion. Firemen have been killed trying this. Always take time to purge a boiler. A furnace explosion is just as deadly as a boiler explosion.

Looking Back

1. The blow down valve is used to dump (empty) a boiler for repairs or inspection, get rid of sludge and mud in the boiler, control high water, and regulate chemical content of the boiler water by removing some of the high mineral water in the boiler.

2. Be sure to keep your hand on the blow down valve when it is open. Operate the valve slowly and watch the water level while you are blowing the boiler.

3. Secure the boiler (shut off the fire) if you cannot see water in the gage glass or if steam comes from the bottom try cock when it is opened.

4. Be sure to have the boiler inspected before operating it again if you had to shut it down for low water. This may prevent an explosion.

5. If any odor of gas or oil is noticed in the firebox shut off all fire lighting devices and check for source of odor. Leaking lines, stuck valves, and defective lighting devices should be looked for.

6. Purge the firebox before attempting to light a fire in gas or oil fired boilers.

PREPARING FOR INSPECTION

Before a boiler can be inspected it must be taken off the line. This has been covered previously in detail. Once the boiler is off the line the following safety checks must be made.

1. The main boiler stop valve or valves (some boilers have

two) must be closed and tagged out. To tag out a valve means to mark it so that it will not be opened by mistake. Some plants have signs that are attached to the valve wheel that read *"Danger! Man in boiler, do not open."*

2. The air cock or top try cock should be checked to see that it is open. This will insure that there is no vacuum inside.

3. The feed water line to the boiler must be closed and also tagged out. If there is an automatic city water make up valve it too must be secured.

The boiler is now allowed to cool slowly. When it is cool enough to dump use the bottom blow down valve to empty the boiler. After the boiler is empty close and tag out the boiler bottom blow down valve.

Note: You can tell if a boiler is cool enough to dump when the heating surface is cool enough to hold your hand on comfortably. Never dump a boiler that is hot. If you do all the sludge and sediment may bake on the heating surfaces. This makes it very difficult to remove.

As soon as the boiler has been dumped open the hand holes or wash out plugs, remove the manhole cover, and thoroughly flush and wash out the water side.

Do not dump a boiler unless you can wash it out immediately. If you dump a boiler and don't flush it right away the sludge and sediment will air dry on the heating surfaces making it extremely difficult to clean.

As you clean the water side look for signs of scale, pitting, or oil. After cleaning the water side you must thoroughly clean the fire side. Remove all soot and carefully examine the entire fire side. Look for signs of blisters or bags on the heating surface. Check the condition of the brick work and make any needed repairs. Also ask the local boiler inspector just what he expects you to have ready. Very often he will want all of the plugs removed at the water column, and low water cut-off controls opened so he can inspect the inside float chamber. If you have a fusible plug it must be replaced. After you have cleaned the boiler, both fire and water sides, notify the inspector that the boiler is ready for him.

Inspection dates are usually the same every year. It would be wise to contact the inspector a couple of weeks in advance and set a date when you will have the boiler or boilers ready for him. This will prevent having your boilers down for an unnecessary period of time.

When the inspector is in your plant have anything ready that he may need, for example, a ladder and a drop light. Have someone to assist him, if necessary. This will save time for both of you. If you notice anything that might affect the safety of the operation of the boiler, call it to his attention. Follow any recommendations the inspector makes. They are for your safety. If a boiler fails he will not be in the building but you will!

HYDROSTATIC TEST

A hydrostatic test is a water pressure test used on a boiler to check for leaks or damage due to a low water condition or put on after any extensive repairs. The boiler inspector may ask for a hydrostatic test if he has any doubts about the boiler being able to carry its rated pressure safely after he has completed his inspection.

In order to put a hydrostatic test on a boiler it has to be completely filled with water. To do this you will have to take care of the following operations.

1. If there is a whistle valve on the water column it must be removed and plugged.

2. Main steam stop valve must be closed.

3. Safety valve or valves must be removed and blank flanges installed or they must be gagged. (A gag is a clamp that will prevent the valve from popping open without damaging the valve.)

4. The air cock must remain open until water comes out. Then close it.

5. Pressure on the boiler is now brought up to 1½ times the maximum allowable working pressure. (The pressure must be

under control so that it does not exceed this pressure by more than 10 lbs.)

Watch the water temperature when filling the boiler. Water that is too cold may cause the boiler to "sweat" and leaks will be hard to find. Water that is very hot may flash into steam when it enters the boiler. After completing the test satisfactorily the whistle valve is replaced and the safety valve gags removed or the valves replaced.

Looking Back ────────────────────────────────

1. Tag out a boiler being inspected on steam lines, water lines, and fire controls.
2. Have all tools and equipment ready for the inspector—drop light, ladders, tools.
3. Testing a boiler for leaks and strength by water pressure requires filling the boiler with water to the top and then applying pressure to the water. Be sure all valves are closed and fastened so the water cannot get into the steam lines or damage the relief valves.
4. Be sure the water is at room temperature or slightly above when filling the boiler for a test. This will avoid sweating due to cold water that could hide a real leak.
5. After a test recheck all valves and return them to their normal operating condition.

────────────────────────────────

LAYING UP A BOILER

Boilers that are to be out of service for any length of time should be laid up to prevent their deterioration. There are two methods of laying up a boiler, wet or dry. Before explaining each in detail it must be understood that regardless of method used the boiler must be thoroughly cleaned on both fire and water sides. It is important that all traces of soot be removed and in a coal fired boiler that all traces of coal and ash be removed. Coal, ash, or soot contain sulphur. The sulphur and water will react to form sulphuric acid which will attack the boiler metal.

Laying up wet is done after the boiler has been thoroughly cleaned on both fire and water sides. Then it should be closed up. New gaskets are used on all hand holes and on the manhole. Be sure that nothing is left inside after the cleaning. Then the boiler is filled to the top with warm chemically treated water to cut down on corrosion and oxygen pitting.

Laying up dry is recommended by the ASME Code for boilers that will be out of service for a long time or if there is any danger of freezing. After the boiler has been thoroughly cleaned, both fire and water sides, the water side must be carefully dried. Any moisture left on the metal surfaces would cause corrosion. The hand holes are replaced using new gaskets. Trays of quick lime, 2 lbs per 1000 gal capacity, or silica gel, 10 lbs per 1000 gal capacity of the boiler, should be placed on the water side of the boiler. Then the manhole cover is replaced, again using a new gasket. All valves and connections are closed to prevent any moisture from entering. The trays of chemicals should be examined at regular intervals and replaced when necessary.

BROKEN GAGE GLASS

To replace a broken gage glass with the boiler under pressure you proceed to replace it by following these steps in order:

1. Secure (close) the water valve to the gage glass.
2. Secure the steam valve to the gage glass.
3. Open the gage glass blow down valve.
4. Remove the gage glass nuts.
5. Remove the broken glass and the washers.
6. Get a new gage glass and new washers.
7. Put the gage glass and the washers in place.

8. Take up on the gage glass nuts hand tight and about ¼ turn with a wrench.

9. Crack steam valve to the gage glass to allow the glass to warm up slowly.

10. Then open the steam valve completely.

11. Now open the water valve to the gage glass. Open completely.

12. Close gage glass blow down valve and check for leaks.

If a gage glass must be cut to fit allow ¼ inch under the inside measurement to allow the glass to expand when it warms up. Also it is important to use new gage glass washers when replacing the glass.

CLEANING GAGE GLASS

There are times when a gage glass becomes dirty on the inside. It is not necessary to replace the gage glass but only to clean it. Here are the steps to follow.

1. Secure water and steam valves to the gage glass.

2. Open gage glass blow down valve and make sure the valves are not leaking steam or water.

3. Remove gage glass nuts and gage glass.

4. Use a cloth wrapped around a wooden dowel to clean the inside. Never use a wire or anything that will scratch the glass. A scratch on the inside of the glass will start the steam cutting the glass and it will break.

5. Use new gage glass washers when replacing the glass.

6. Follow the warm up procedures given in points 8-12 under Broken Gage Glass.

LEAKING GAGE GLASS WASHERS

A leaking gage glass will give a false water level reading. Also the gage glass will wear and eventually break. Take up on the gage glass washers by tightening the gage glass nuts carefully. If tightening the nuts does not stop the leaking then replace the washers following the procedure under Broken

Gage Glass. Wear safety goggles when working around the gage glass and also secure the steam and water valves to the gage glass. Keep a supply of gage glass washers on hand.

CORRECTING A STEAMBOUND PUMP

A pump becomes steambound when the water it is pumping becomes too hot and turns to steam. As the pump is designed to handle liquids it will not work with steam or air. The usual cause of the trouble is a faulty steam trap that is allowing steam to blow into the return lines. If the condition isn't corrected promptly it could damage the pump or the motor driving it.

To correct the steambound condition the water coming to the pump must be cooled. If the pump is taking its suction from a receiver or feed water heater city water can be added through the make-up system to drop the temperature. Then the traps returning water to the tank or receiver should be checked and repaired. See Chapter 4.

A vacuum pump that is getting back condensate that is too hot will make a loud banging noise. It sounds like a lot of nuts and bolts banging around inside. Cool the water by adding city water through the make-up system. If this doesn't correct the situation and cause the pump to start pumping again then run cold water over the pump. Be sure to run the water over the pump, never on the motor.

TESTING LOW WATER CUT-OFF

There are two ways to test the low water cut-off. One is with the burner firing. Open the blow down valve on the low water cut-off. The water and steam rushing out will clean out any sludge and will also allow the float to fall. This will shut off the fuel valve just as if the boiler was in a low water condition. This should be done daily on every shift.

The second way is to test the low water cut-off by actually causing a low water condition to occur. You can make this test by securing the vacuum pump or the feed water pump so no water is supplied to the boiler and the water level will gradually drop. This is a more accurate test than with the low water cut-off blow down valve. Under normal conditions the water level drops slowly and there is a possibility of the valve sticking. When a blow down is used to test the valve the rush of water and steam would jar the float and make it drop. The burner should shut off when water is still visible in the gage glass. If the low water cut-off fails to cut off the burner then have someone in attendance with the boiler as long as it is firing and until it can be taken off the line and checked out.

Looking Back

1. Remove all soot from fireside of boiler when laying it up. Clean all surfaces carefully to remove all traces of ash, fuel, or soot as these can corrode.

2. Fill the boiler to the top with warm, treated, deaerated water and close all valves when laying up for a short time.

3. For a long lay up clean out all sediment and scale, drain and dry the water side. Add trays of moisture absorbent to keep the air dry. Close up all openings and close all valves.

4. Gage glasses are extremely important. Learn how to replace a broken glass or to clean one and to prevent the washers from leaking.

5. Return water pumps may become steambound if extremely hot water and steam come to them in the condensate return lines. Cool the condensate water by adding city water to it. Be sure to find the source of the steam getting into the return lines and correct the leak.

6. Test low water cut-off every day by opening the blow down valve on the cut-off. This will drop the level in the cut-off and cause it to operate to cut off the fuel supply. Test actual low water conditions monthly.

Who was to blame? The boiler or the fireman?

9

BOILER ROOM SAFETY

Safety is always in season and never happens just by chance. It is achieved by looking around and ahead and being aware of the things that could cause accidents. A good set of safety rules will list many of the possible causes of accidents and ways to avoid them. The rules given here are ones based on boiler rooms and you would do well to check them over carefully. You may wish to add some of your own that apply to your particular plant.

REPORTING ACCIDENTS

Even though shop safety rules are followed conscientiously there is always the possibility of an accident happening. Be sure to find out the procedure to be used in your plant. All accidents should be reported and learn how to report them properly. Always have all cuts and splinters given first aid. In many plants failure to report accidents and injuries can cause trouble with insurance companies, especially if complications arise later.

FIRE PREVENTION

In the boiler room we use fire and fuel to produce steam. There is always the chance that some accident may cause the fire to occur outside the boiler. The fireman is concerned with putting out a fire in the boiler room as well as in the boiler. He should know the causes of the fire and the ways of controlling it. He should be familiar with the different types of fire extinguishers to use.

There are three things needed to start a fire:

1. Fuel or something that will burn.
2. Heat enough to cause the fuel to start burning.
3. Air to supply oxygen for the fire.

CLASSES OF FIRES

Fire officials classify fires into three common classes according to the material that is serving as fuel for the fire.

Class A are fires in wood, paper, coal, rubbish, buildings, and most common materials.

Class B fires are in liquids such as oil, gasoline, paint, fats, and similar materials that create a gas when heated.

Class C fires are in live electrical equipment such as motors, switches, fuse boxes, and appliances.

If there are three things needed to start a fire, then there are three ways that can be used to stop one. If we eliminate one of the necessary items for a fire then the fire stops. So the three ways to stop a fire are:

1. Cut off the fuel supply. When the fuel is burned up the fire goes out.

2. Cool the burning material so it is below the ignition temperature. This is why water is used on many fires. The water cools off the fuel rapidly.

3. Cut off the air supply so that no oxygen can get to the fuel. Smothering a fire by covering it with sand, soil, or various kinds of chemicals does this.

FIRE EXTINGUISHERS

There is no single agent that will put out all kinds of fires. The fire extinguishers that are placed around buildings are called first aid fire extinguishers. They are not intended to take the place of a fire department. They are meant to put out small fires or to slow down larger fires until the fire department arrives. You should know that a fire can multiply itself fifty times in eight minutes and with a class B fire even seconds are important.

Find out the procedure to be used in sending for the fire department or sounding a fire alarm. Make sure someone can direct the fire department to the right location when you send for them. Are there fire alarm boxes or stations in or near the boiler room? Learn the location and the use of them.

The types of the commoner first aid fire extinguishers are shown in Fig. 9-1. Note that each one is for a certain type of fire. It is important that the right type be used on the fire. For example, water used on a class B fire can cause it to spread. Water used on a class C fire can cause short circuiting and start additional fires or it can electrocute you instantly.

Look around your boiler room. Make sure that you have the proper types of extinguishers and know how to use them.

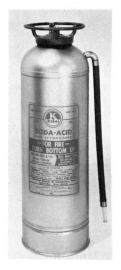  

Fig. 9-1. Types of fire extinguishers, water (soda-acid), foam, carbon dioxide. Walter Kidde Co.

In some cities a law requires that buckets of sand be kept near each boiler. The sand is useful for controlling an oil fire. Spreading sand over the oil fire will smother it. It is also useful in event of an oil spill. The sand will absorb the oil and prevent it from spreading. It will also prevent slipping on the oil.

Check with your local fire department. Some departments have trained men (Fire Prevention Bureau) that inspect buildings and factories and suggest ways to prevent fires. Many insurance companies have inspection services. All of them will be glad to give you any information or help that they can.

HOUSEKEEPING

Good housekeeping goes hand in hand with safety. It's easy to keep a boiler room clean when it's a daily practice. A few small investments in time and money may save millions in damage repairs. Oily rags can lead to a fire due to spontaneous combustion. Oil on the floor can cause a fire or a serious accident. Safety cans for combustible liquids may prevent a flash fire. Using gasoline for cleaning is dangerous and foolish. A cup of gasoline has enough force to destroy a single car garage if it is turned into an explosive gas. A few rules to remember in your daily housekeeping are listed.

1. Store oily rags in approved containers.
2. Store all combustible liquids in safety cans.
3. Wipe up all oil spills at once.
4. Never use gasoline, naphtha, or similar materials for cleaning.
5. Do not use carbon tetrachloride for cleaning.

SAFETY RULES

The following safety rules were published by the magazine *Power Engineering* in its issue of January 1965. The rules are for people working with boilers and power plants. Study these carefully and you may be able to avoid many accidents.

FOLLOW THESE SAFETY RULES

A better accident history can only result from the combined efforts of all. Carelessness accounts for the majority of accidents. Go over this *Never List* and see how many apply to your job.

1. Never wear frayed clothing which might catch in machinery or cause tripping.

2. Never wear shoes that could cause slipping, or with worn soles that permit easy puncture.

3. Never fail to wear a helmet-type hat where there is danger of head injury from electric wiring or falling objects.

4. Never expose your eyes to flying particles or the extended ray of a welding machine.

5. Never leave your hands unprotected from burning, slivers, or abrasions.

6. Never use a flashlight unless it gives suitable light and is dependable, so as not to leave you in the dark in some remote area of a large vessel or boiler.

7. Never use an extension light with a weak or frayed cord, or without a cage around the bulb.

8. Never experiment with electrical *hook-ups* beyond your knowledge.

9. Never take a step in a dark area if you can't see where you are putting your foot down.

10. Never enter the furnace of an automatically-fired boiler without making sure that the firing mechanism is in the *locked out* position and is tagged to this effect. Never enter the furnace of a water-tube boiler without watching for falling slag.

11. Never enter a boiler installed in battery without making sure that all stop valves in the connecting pressure lines are securely closed and tagged. If there are bleed-off connections between the stop valves, open them.

12. Never drain a vessel or boiler without opening a suitable vent. This will prevent collapse and permit complete drainage.

13. Never remove a handhole or manhole cover plate from a vessel or boiler unless you are sure that it is empty.

14. Never enter a boiler or pressure vessel unless a reliable person is stationed nearby to note any personal mishap or weakness.

15. Never use a ladder without checking for worn rungs and loose side rails.

16. Never use a ladder that is too short. In trying to reach up, the angle of the ladder and user becomes such as to cause falling backward. When trying to enter an inspection opening, the ladder should extend to or above it. Climbing from the uppermost rung can cause the ladder to tip out at the top or kick out at the bottom. In placement watch out for electric wiring.

17. Never use a ladder not properly adapted to the job. The bottom ends of the rails should be fitted with spikes or cleats to prevent slipping. If a ladder is not so equipped, or is placed on a metal floor, secure the bottom to some fixed object. Do not rely on someone to hold the ladder, as he may be called away on another mission or emergency.

18. Never risk your life by removing fuses from high-voltage electrical circuits with your bare hands. Use tongs or fuse pullers.

19. Never use ladders as bridges.

20. Never enter a pressure vessel that contained a toxic agent or may contain *dead air* without thorough purging. If necessary, wash and clean the surfaces.

21. Never use a rope ladder or bosun's chair without first inspecting the rope. Also, be sure that you are physically able to make an inspection from such flexible equipment.

22. Never inspect a boiler or pressure vessel while it is under hydrostatic pressure test without using due care. Always make sure a vessel is completely vented of air before a hydrostatic test is applied.

23. Never operate soot blowers unless the burners are operating at a high firing rate. Air flow should be adjusted to insure a high carbon dioxide but low oxygen content in the flue gases.

24. Never enter a building or other enclosure in the pres-

ence of ammonia fumes, sewer gas, natural gas or any other toxic gas without proper clothing and respiratory equipment— and then only with extreme caution.

25. Never at any time stand in front of the discharge opening of a relief valve—particularly while attempting to test it.

26. Never blow down an appliance on a boiler without first observing the point of discharge—your foot may be under it.

27. Never blow down a boiler under pressure by opening the slow-opening valve first. Always open the cock first and then the slow-opening valve. After blowdown close the slow-opening valve first. This prevents a sudden back pressure wave on the connecting piping.

28. Never shut a valve off suddenly in a pressure line except under extreme emergency. The sudden change in flow may result in rupture at some weakened location. Also, never open a valve suddenly (particularly in a steam line) where water hammer might result.

29. Never operate a pressure vessel with only part of the head or cover plates, bolts, lugs or clamps in place.

30. Never continue to use any bolt, pin, lug, or clamp after it becomes worn, sprung, stripped or otherwise weakened.

31. Never release the holding mechanism of any quick-opening door or end closure until sure that the chamber is void of pressure.

32. Never depend on automatic controls as safety devices unless they are kept clean and are checked, tested and otherwise regularly maintained.

33. Never operate boilers, pressure vessels or machines at pressures or speeds above rated.

34. Never take someone else's word for something you should have checked yourself.

35. Never resort to *horse-play*.

36. Never neglect your fire protection. Check fire buckets and extinguishers regularly, and never leave a water line valve shut off to a sprinkler system through lack of maintenance. Red tag it and stay with it until corrections or alterations have been completed; and then return all valves to their proper positions for service.

Looking Back ——————————————————————

1. Report all accidents and have all injuries, no matter how small, treated.
2. There are four types of fire extinguishers—water, foam, carbon dioxide, and dry chemical. Locate the ones in your boiler room and learn their types and uses.
3. A clean boiler room will have fewer accidents as many causes will be removed.
4. Learn the procedure in your plant for calling the fire department or sounding a fire alarm.
5. Study the safety rules and learn to use them. These rules are based on actual cases and many of them may not occur to you.

REVIEW QUESTIONS

Listed below are a series of questions that every fireman should be able to answer. They will prove to the fireman that he understands his plant and how to operate it safely. These questions are the type of questions that would be asked by examiners in the states and cities that require the operator of a low pressure boiler to carry a fireman license.

1. What is the first thing you do upon entering the boiler room?
2. What would you do if water came out when you opened the top try cock?
3. What would you do if steam came out when you opened the bottom try cock?
4. How many ways do you have of getting water into your boiler?
5. What is a pressuretrol? How many do you have on your boiler?
6. How often do you test your safety valves? How do you test them?
7. Where is a check valve found on the boiler and what purpose does it serve?
8. How much steam do you carry on your boiler? What is your operating range?
9. What is the purpose of a low water cut off? Where is it located? How often do you test it? How do you test it?
10. Where does the bottom blow down line discharge to? Why is this necessary?
11. What are your safety valves set to pop at?
12. How would you prepare a boiler for inspection?
13. What is meant by purging a boiler? How would you purge a boiler?
14. How would you replace a broken gage glass with the boiler under pressure?
15. Can you run a boiler with a broken gage glass?
16. What is the purpose of a fire-eye?
17. How do you test the fire-eye? How often do you test it?
18. What is the most important valve on the boiler? Why is it the most important valve?
19. How many valves are there on the bottom blow down line of a boiler?
20. How often do you blow down a boiler?
21. What happens when you blow down a low water cut-off?
22. What would you do if a low water cut-off failed to shut the boiler off when testing?
23. Do you have pressure on the boiler when you blow it down?
24. Give the following information on your boiler: (a) Name and type of boiler; (b) Horsepower size; (c) Steam pressure carried; (d) Safety valve setting.
25. How many ways do you have of finding the water level in a boiler?
26. How often do you blow down the following: gage glass, water column and low water cut-off?
27. What is smoke?
28. How many kinds of draft do you find in a boiler?

29. How do you know that it is safe to open the boiler up for inspection?
30. What prevents the water from the boiler backing up into the feed water line when the feed water pump stops?
31. Why do you need air in the boiler room?
32. What would you do if upon entering the boiler room you found the safety valve popping and the pressure gage showed 30 pounds pressure on the boiler?
33. What procedure would you follow if you found your boiler off on safety when you first entered the boiler room?
34. Name some of the automatic controls on your boiler.
35. How many valves are found on the feed water line between the boiler and the water source?
36. What is complete combustion?
37. Where does the feed water line enter your boiler?
38. What would you do if upon entering the boiler room you found the gage glass full of water?
39. How would you go about closing up a boiler and put it back on the line after inspection?
40. What is a furnace explosion? What precautions would you take to prevent a furnace explosion?

GLOSSARY

Air Cock—Valve located on line coming from highest part of steam side of the boiler. It is used to vent air from the boiler when filling; when warming up; and to allow air into boiler when taking it off the line to prevent a vacuum.

ASME Code—Code adopted by the American Society of Mechanical Engineers for construction and operation of boilers and related equipment. This code has been adopted by most of the 50 states.

Atomization—The breaking up of oil in small particles (mist) so that it may be surrounded by air for better combustion.

Automatic City Water Make-Up—Sometimes referred to as a water feeder. It is located a little below the normal operating water level and used to add city water directly to the boiler in event of failure of return water. It is not to be used as a feed water regulator.

Blow Down Valve—Found on lowest part of water side of fire tube boiler and on mud drum of water tube boiler. Usually found in pairs, used to: control high water, control chemical concentrations, remove sludge and sediment, and dump a boiler for cleaning.

Boiler, Cast Iron Sectional—Boiler built of cast iron sections connected with push nipples and held together with through stays.

Boiler, Fire Tube—Boiler that has heat and gases of combustion pass through tubes surrounded by water.

Boiler, Water Tube—Boiler that has water inside tubes with heat and gases of combustion passing around the tubes.

Btu—British Thermal Unit—a measurement of heat. The amount of heat needed to raise the temperature of one pound of water one degree fahrenheit.

Carryover—Slugs of water being carried over into the steam line with the steam. This leads to water hammer and possible pipe rupture.

Chemical Feeders—Located on the discharge side of a feed water pump or vacuum pump. Used to add chemicals to treat the boiler feed water.

Combustion—The rapid burning of fuel and oxygen.

Combustion Controls—Automatic devices that control the efficient combustion of fuel. They regulate the starting and stopping of the burner, the high and low fire of the burner, the purging of the fire box, the secondary and primary air.

Complete Combustion—The burning of all the fuel with the proper amount of excess air.

Condensate—Steam that has been condensed back into water and eventually returned to the boiler.

Draft—A difference in the pressure between 2 points usually the atmosphere and the boiler setting, may be forced, natural, or induced, or a combination of above. Needed in order to burn fuel. It takes about 15 pounds of air to burn one pound of fuel.

Draft Gage—Device used to record draft. It measures a difference in pressure between the atmosphere and the boiler fire box or the breeching, or the stack or in some plants all three. Calibrated in tenths of inches of water.

Erosion—Wearing away of metal due to wet steam leaking through a valve.

Fast Gage—A pressure gage that records a higher pressure than is actually present.

Feed Stop & Check Valve—Located on feed water line as close to the boiler as practical. Stop valve is closest to the shell of the boiler. Check valve prevents water from backing out of boiler. Stop valve allows check valve to be repaired without dumping the boiler.

Feed Water Heater—Condensate returns here. Water is heated and air and other non-condensable gases are vented to the atmosphere. It is located on the suction side of the feed water pump above the feed water pump.

Feed Water Lines—Those lines from the feed water pump to the boiler. They return the necessary water to the boiler. The fireman should know where they are and every valve to be found on them.

Feed Water Regulator—Located at the normal operating water level. Sometimes referred to as a pump control. Used to maintain a constant water level in the boiler.

Fire Eye—Device used to prove pilot and main flame on a boiler. In event of pilot failure or main flame failure it shuts off the fuel supply to the burner, preventing fuel from flooding the fire box and causing a furnace explosion.

Fire Point—The temperature at which oil will burn continuously when exposed to an open flame.

Flash Point—The temperature at which oil, when heated, will produce a vapor that will flash when exposed to an open flame.

Foaming—Rapid fluctuation of water level due to high surface tension or high total dissolved solids.

Forced Draft—Mechanical draft that forces air into the firebox under a slight pressure.

Fuel Oil Burners—Used to atomize the oil for efficient combustion.

Fuel Oil Pump—Takes the oil from the fuel oil tank and delivers it to the burner under pressure.

Fuel Oil Strainers—Usually of duplex type found on fuel oil suction lines before the fuel oil pump. Allows one strainer to be cleaned while the other is in operation without having to shut off boiler.

Fuel Oil Thermometers—Found on various parts of the fuel oil system to record the temperature of the fuel oil that fireman must know in order to efficiently burn fuel.

Fusible Plug—Last warning device before burning up heating surface. Located 1″ to 2″ above the highest heating surface. Brass or bronze plug with a tapered hollow center filled with 99% pure virgin banca tin—melts at about 450°F.

Gage Glass—Located on water column used to show how much water is in the boiler.

Gage Glass Blow Down Valve—Located on bottom of gage glass to check the water level in the boiler and blow out any build up of sludge and sediment and prove the steam and water lines are clear.

Gate Valve—Found on lines in the boiler room. Used wherever a direct flow through the valve with no drop in pressure is required. Especially found on main steam lines and on main header.

Globe Valve—Found on lines in the boiler room where a throttling action is required. It is designed for this and the seats are easily repaired or replaced.

Hard Water—Water that contains large quantities of minerals and scale forming salts.

Heating Surface—Any part of the boiler that has water on one side and heat and gases of combustion on the other side where there can be a transfer of heat.

High Fire—Position of burner control at which it will burn maximum amount of fuel.

High Surface Tension—Impurities that are floating on the surface of the water level in the boiler. Dangerous because it leads to foaming, priming, and carryover.

High Water—When water starts to go well over the normal operating water level. This is dangerous because it could lead to carryover.

Ignition Transformer—Used to supply a spark that ignites the gas pilot that in turn ignites the fuel oil.

Incomplete Combustion—Condition which exists when all the fuel is not burned. The result is smoke and soot.

Induced Draft—Mechanical draft that causes air to enter the firebox of a boiler by causing a vacuum in the firebox and the air moves in to take its place.

Lead Sulfite Cell—Detector that is sighted to pick up pilot and main flame and pass signal on to programming clock.

Low Fire—Position of burner control at which it will burn minimum amount of oil. All burners should start up and shut off in low fire.

Low Water Cut-off—Sometimes referred to as a Low Water Fuel Cut-Off. Located a little below normal operating water level. It will shut off the boiler burner in the event of low water preventing burning out of tubes and possible boiler explosion.

Main Steam Stop Valve—Located on main steam outlet from boiler. Used to cut boiler on or off the line usually an O. S. & Y. gate valve.

Modulating Pressuretrol—Located next to pressuretrol. Controls boiler high and low fire position through the modulating motor.

Natural Draft—Depends on the height of a chimney, and the difference in weight, between a column of warm air and an equal column of cold air.

Normal Operating Water Level—Usually about ½ a gage glass, or about 6″ above the highest heating surface.

O. S. & Y. Gate Valve—Outside stem and yoke, sometimes referred to as a rising stem valve. It shows by its position whether it is open or closed.

Perfect Combustion—The burning of all the fuel with only the theoretical amount of air.

Pitting—Concentrated oxygen attack on boiler metal. Somewhat like a woodpecker making a hole in a tree.

Pressuretrol—Automatic device located on line coming off highest part of steam side of boiler. Controls operating range of boiler. Starts and stops boiler on pressure demand.

Priming—Small particles of water being carried over with the steam.

Primary Air—Air supplied to the burner that regulates the rate of combustion.

Programming Clock—Controls the firing cycle of a boiler, starting with purging a boiler, pilot, main flame and post purge.

Purge a Firebox—To get rid or drive out the fumes in a firebox that could lead to a furnace explosion.

Return Lines—Those lines that carry the condensate back to boiler room.

Safety Valve—Found on highest part of steam side of boiler. Prevents the boiler from exceeding its maximum allowable working pressure by relieving excess steam pressure.

Secondary Air—Air supplied to the burner which controls the efficiency of combustion.

Siphon—Located on steam gages to prevent live steam from entering and warping Bourdon tube. Also used to protect pressuretrols.

Slow Gage—A pressure gage that records a lower pressure than is actually present.

Soft Water—Water that contains none or only small traces of minerals and scale forming salts.

Solenoid Valve—Valve that is opened electrically.

Steam Gage—Connects to highest part of steam side of boiler. Records steam pressure carried on the boiler. It is read in pounds per square inch (psi).

Steam Gage Compound—Same location as steam gage. Differs in that it records both pressure and vacuum. Pressure in psi and vacuum in inches.

Steam Lines—Used to take the live steam to where it will do its work.

Steam Traps—Automatic device that increases the overall efficiency of a plant by removing air and water without the loss of steam. Found wherever steam is used to give up its heat. Also on main headers and branch lines.

Strainers—Used to trap scale and impurities that might clog up a trap. Found on lines before inverted bucket and float type traps especially.

Surface Blow Down Line—Located at the normal operating water level. Used to remove impurities from the surface of the water that could cause surface tension.

Temperature Regulators—Automatic device used to regulate the steam flow to various heat exchangers (fuel oil tanks, hot water tanks, etc.).

Total Dissolved Solids—Solids that have built up in the water side of the boiler. Must be reduced by giving the boiler bottom blows.

Try Cocks—Usually three located on water column, a second way of determining the water level. Used as a check against the gage glass and also when gage glass is being replaced or cleaned.

Vacuum—An absence of pressure, a pressure below atmospheric pressure measured in inches of water.

Vacuum Pump—Found on heating systems, causes positive return of condensate to vacuum tank. Designed to handle air and water. Air is discharged to atmosphere and water either to boiler, receiver or feed water heater.

Water Column—Found at normal operating water level, casting on which try cocks and gage glass are mounted. Used to slow down fluctuation of boiler water to get a better reading on gage glass.

Water Column Blow Down Valve—Located on bottom of water column. Used to blow out any build up of sludge and sediment.

Wire Drawing—Hair line cutting of metal due to live steam leaking through a valve.

INDEX

Illustrations are listed in italic.